State and Provincial Amphibian and Reptile Publications For the United States and Canada

Wahlgreniana 3

State and Provincial Amphibian and Reptile Publications For the United States and Canada
Second Edition

John J. Moriarty & Aaron M. Bauer

International Society for the History
and Bibliography of Herpetology
www.ISHBH.com

2024

Wahlgreniana is named in honor of Richard Wahlgren (1946–2019) A founding member and first Chairman of the *International Society for the History and Bibliography of Herpetology*. Without Richard's tireless dedication to ISHBH, the society could not have made it through its early years. *Wahlgreniana* is a series of book length works complementing the ISHBH journal, *Bibliotheca Herpetologica.* Books in this series are published on an irregular basis and are sold separately from ISHBH subscriptions.

International Society for the History and Bibliography of Herpetology
© 2024 ISHBH & John J. Moriaty & Aaron M. Bauer
All rights reserved. Published 10 June 2024
Printed on demand by IngramSpark global distribution network
Layout & design by Breck Bartholomew

ISBN: 979-8-218-44771-7

www.ISHBH.com

Cover and titlepage illustration: Muhlenberg's (Bog) Turtle (*Clemmys muhlenbergii*). Plate 28 *in* Babcock, Harold L. 1919. The Turtles of New England. Memoirs of the Boston Society of Natural History 8:325–431, pls. 17–32.

Table of Contents

Introduction

This second edition bibliography is intended as a guide to the regional (state and provincial) herpetofaunal literature of the United States and Canada through December 31, 2023. We have added new publications since 1999, plus older publications that were missed or excluded in the first edition (Moriarty and Bauer. 2000. State and Provincial Amphibian and Reptile Publications for the United States and Canada. *SSAR Herpetological Circular 28*. 52 pp.)

This literature had its beginnings in the early nineteenth century with catalogues and summaries of the herpetofaunas of the northeastern states (e.g., Storer 1839 for Massachusetts, De Kay 1842 for New York) and expanded westward with "civilization." Likewise, progress in Canadian regional treatments began in the Maritimes (e.g., Gilpin 1875 for Nova Scotia, Cox 1898, 1899 in New Brunswick) and expanded toward the Prairie Provinces. A glance through this bibliography will indicate that the work is not yet complete. New descriptions and range extensions as well as taxonomic changes have rendered many classic regional works obsolete (e.g., Fowler 1907 for New Jersey) and in some areas thorough treatments have never been published.

This edition includes 1323 references (Table 1), 440 more citations than the first edition. Forty-eight citations are listed for more than one state for a total of 1275 unique listings. We are aware that the list we have compiled is incomplete. Nonetheless, we are confident that we have identified the majority of the significant regional contributions to the North American herpetofauna and trust that users will be able to trace even more obscure references, as well as more locally focused publications (county herpetofaunas, etc.) through the citations provided here. This should be facilitated by our inclusion of state/provincial herpetofaunal bibliographies, as well as checklists, keys, and more extensive treatments.

We have limited our scope to include those published papers and books that deal primarily with single states or provinces or substantial portions thereof. Regional works of broader scope, covering up to three states or provinces have also been included. Only in cases of New England (Connecticut, Maine, Massachusetts, New Hampshire, Rhode Island, and Vermont) and the Maritimes (New Brunswick, Newfoundland, Nova Scotia, and Prince Edward Island) have more inclusive areas been admitted. In both cases herpetofaunal diversity is relatively low and the constituent political regions have tended to be discussed as a group. Most publications dealing with smaller areas, such as counties, have not generally been included. Some in depth county reports (i.e., longer than ten pages) are included, since they normally contain information beyond the county boundaries. However, works dealing with large parks, significant drainage basins, mountain chains, and other subcomponents of individual states have been included on a case-by-case basis. We did include shorter articles for four authors: Bumpus (Rhode

Island), Harrison (Pensylvania), Reed (Delmarva), and Strecker (Texas). Their collective papers create an early complete corpus for their respective regions. Although at some level our choices are arbitrary, we have attempted to include works on these smaller units if they were representative of most of a state, covered parts of states or provinces that were not well represented by more inclusive works, or were especially comprehensive with respect to the herpetofauna included. Thus a work covering a county or region that was comprehensive was likely to be included, whereas one based on a single year's collecting or one collector's records alone was not. Regional publications dealing comprehensively with ordinal or subordinal level diversity were also included (e.g., The snakes of—, The salamanders of—). Publications on venomous snakes are only included for states with four or more species. Familial treatments on a regional scale were also included if the group involved was not represented by only a single species and the emphasis was on distribution and related issues.

We have included most books and journal articles but have excluded most gray literature. We also included herpetological sections of more general publications that cover the natural history or vertebrates of a state or region, primarily for publications pre-1970. After that date reptile and amphibian specific books were available for all states. We were especially permissive for very early works, which represented one of the first reports of a regional herpetofauna, even if they were a small part of much larger works (e.g., Kirtland 1838 for Ohio or Kennicott 1854 for Cook County, Illinois). Regularly numbered and generally available federal or state reports have been included, but internal reports, mimeographed in-house notes from universities or museums, theses, dissertations, posters, and leaflets (unless part of a regularly published series) have been excluded. We have also excluded websites, laminated folded pocket guides and some self-published guides that were incomplete (with a few exceptions for which no recent conventional publications were available). Although these sources of information become more numerous with every year and may be very useful to some users, they are bibliographic nightmares and, especially in the case of websites, authorship may be uncertain and the "publications" themselves may be ephemeral.

Some older publications (i.e., Storer 1839, 1840—Massachusetts, Hay 1887, 1887, 1892—Indiana) were published in several versions simultaneously, or nearly so. When available to us we listed multiple versions of many entries in square brackets following the main entry, along with other additional information of bibliographic relevance. Revised editions of works as well as reprints have likewise been noted, especially if they contain significant new information and/or if they bear different volume numbers in regularly published series (although many works, especially those issued by state agencies, have been issued in an almost untraceable number of versions, and we make no claim to completeness in this regard). Thus, all of the titles listed should be traceable and ultimately obtainable by interested workers. Two possible exceptions are the publications of the Reed Herpetorium and the Ross Allen Reptile Institute. In both cases these private herpetological concerns produced significant regional publications that we felt could not be excluded. Nonetheless, these publications were widely distributed and there are many copies in herpetological libraries.

We envision that this bibliography will be used by a broad range of herpetological users. While modern search engines are spectacular tools, those looking for specialty literature that is older, more obscure, or of limited regional focus will likely find this bibliography of particular use. In addition to research use, we also hope that it will be of use to collectors of state and provincial literature, who can use it as a checklist to guide the growth of their libraries. In truth, our motivation in compiling this list was, at least in part, to discover what we were missing! Bearing in mind the diversity of potential users, we have attempted to use as uniform citation style as possible, adapting it as necessary for some of the more bibliographically challenging entries. We have spelled out journal titles in full and have given complete pagination for stand-alone works, including pages numbered with Roman numerals and unpaginated or separately numbered plates. This should facilitate the location of these references, but we caution that users of this bibliography should examine copies (physical or at least digital) in order to verify the accuracy of the information. Finally, some citations appear in **green**. These denote entries for which cover images are provided.

Although we compiled this bibliography with the aforementioned criteria for inclusion in mind, our choices were, of necessity, somewhat subjective and many titles were left out that other bibliographers might have included. We also recognize that, despite our best efforts, there are books and articles that we have missed. We would appreciate receiving the complete citation of any omissions and will continue to gather new references to update this list.

We thank our colleagues and fellow bibliophiles who made us aware of many of the titles listed and helped with incomplete citations for one or both editions of the bibliography, especially Kraig Adler, Breck Bartholomew, Chris Bell, Al Breisch, Joe Collins, Drew Davis, Ken Dodd, Bob Hansen, John Levell, Ernie Liner, Joe Mitchell, Floyd Scott, Frank Slavens, Doug Thomson, and Tom Tyning.

We thank Chris Bell and Drew Davis for reviewing an earlier version of this publication and Breck Bartholomew for suggesting merging our supplement with the original edition.

Omitted and new citations should be sent to:

John J. Moriarty
Bell Museum of Natural History
University of Minnesota
2088 Larpenteur Ave W.
St Paul, MN 55113
e-mail: frogs@umn.edu

Aaron M. Bauer
Department of Biology
Villanova University
800 Lancaster Ave.
Villanova, PA 19085–1699
e-mail: aaron.bauer@villanova.edu

Table 1. Number of Citations by State and Province

State	Citations	State	Citations
Alabama	17	North Dakota	6
Alaska	6	Ohio	37
Arizona	36	Oklahoma	31
Arkansas	14	Oregon	11
California	61	Pennsylvania	43
Colorado	23	Rhode Island	10
Connecticut	10	South Carolina	24
Delaware	16	South Dakota	15
Florida	65	Tennessee	24
Georgia	15	Texas	69
Hawaii	12	Utah	21
Idaho	15	Vermont	5
Illinois	33	Virginia	58
Indiana	25	Washington	22
Iowa	12	West Virginia	29
Kansas	41	Wisconsin	29
Kentucky	19	Wyoming	9
Louisiana	26		
Maine	11	**Province**	**Citations**
Maryland and Washington D.C.	34	Alberta	8
Massachusetts	22	British Columbia	15
Michigan	35	Manitoba	6
Minnesota	33	Maritimes	5
Mississippi	16	New Brunswick	3
Missouri	29	Newfoundland	2
Montana	13	NW Territories	5
Nebraska	9	Nova Scotia	8
Nevada	9	Nunavut	4
New England	16	Ontario	27
New Hampshire	6	Prince Edward Is.	4
New Jersey	10	Québec	13
New Mexico	18	Saskatchewan	5
New York	37	Yukon	3
North Carolina	28	**Total***	**1323**

* The actual total is 1275 because some titles are listed for more than one state or province.

Alabama

Boschung, H. 1977. How to know the poisonous snakes of Alabama. Nature Notebook Alabama Museum of Natural History, No. 3. 7 pp.

Chermock, R.L. 1952. A Key to the Amphibians and Reptiles of Alabama. Geological Survey of Alabama, Museum Paper 33. 88 pp.

Gibbons, J W., Haynes, R.R., and Thomas, J.L. 1990. Poisonous Plants and Venomous Animals of Alabama and Adjoining States. University of Alabama Press, Tuscaloosa, AL. xv + 345 pp. [amphibians and reptiles on pp. 302–316, 321–323, pls. 23–24].

Guyer, C., M.A. Bailey and R.H. Mount. 2015. **Turtles of Alabama**. University of Alabama Press, Tuscaloosa, AL. xv + 267 pp.

Guyer, C., M.A. Bailey, and R.H. Mount. 2018. Lizards and Snakes of Alabama University of Alabama Press, Tuscaloosa, AL. xvii + 397 pp.

Guyer, C. and M.A. Bailey. 2023. Frogs and Toads of Alabama. University of Alabama Press, Tuscaloosa, AL. xii + 247 pp.

Haltom, W. L. 1931. Alabama Reptiles. Geological Survey of Alabama, Museum Paper 11. vi + 145 pp.

Linzey, D.W. 1972. Snakes of Alabama. Strode Publishers, Huntsville, AL. 136 pp.

Löding, H.P. 1922. A Preliminary Catalogue of Alabama Amphibians and Reptiles. Geological Survey of Alabama, Museum Paper 5. 59 pp.

Mirarchi, R.E. (ed.), 2004. Alabama Wildlife. Volume One. A checklist of Vertebrates and Selected Invertebrates: Aquatic Mollusks, Fishes, Amphibians, Reptiles, Birds, and Mammals. The University of Alabama Press, Tuscaloosa. viii + 209 pp. [amphibians and reptiles on pp. 101–132].

Mirarchi, R.E., Bailey, M.A., Haggerty, T.M. and Best, T.L. (eds.), 2004. Alabama Wildlife. Volume Three. Imperiled Amphibians, Reptiles, Birds, and Mammals. The University of Alabama Press, Tuscaloosa. x + 225 pp. [amphibians and reptiles on pp. 9–96].

Mount, R.H. 1975. **The Reptiles and Amphibians of Alabama**. Auburn University Press, Auburn, AL. vii + 347 pp.

Mount, R.H. 1976. Amphibians and Reptiles. pp. 66–79 *in* H. Boschung (ed.) Endangered and Threatened Plants and Animals of Alabama. Bulletin of Alabama Museum of Natural History 2:1–92.

Mount, R.H. 1984. Vertebrate Wildlife of Alabama. Alabama Agricultural Experiment Station, Auburn University, Auburn, AL. 44 pp. [amphibians and reptiles on pp. 15–24].

Mount, R.H. 1986. Vertebrate Animals of Alabama in Need of Special Attention. Auburn University Agricultural Experiment Station, Auburn, AL. 124 pp.

Shelton–Nix, E. (ed.), 2017. Alabama Wildlife. Volume Five. The University of Alabama Press, Tuscaloosa. xvi + 355 pp.

Wimberly, C.A. 1970. Poisonous Snakes of Alabama. Explorer Books, Inc., Birmingham, AL. 46 pp. [amphibians and reptiles on pp. 191–255].

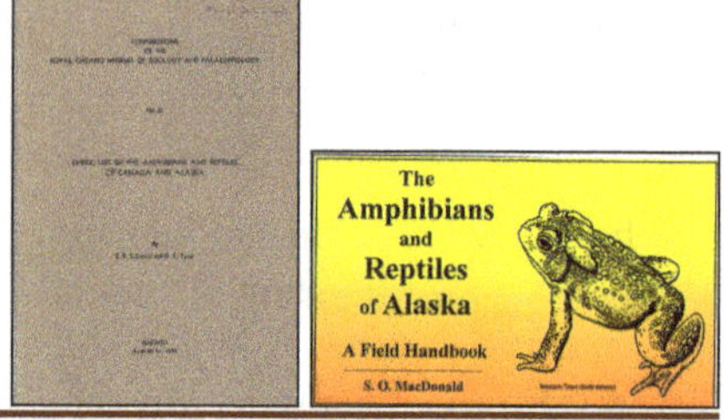

Alaska

Hock, R.J. 1967. Alaska zoogeography and Alaskan Amphibia. pp. 201–206 *in* Proceedings of the Fourth Alaska Science Conference of 1953, Anchorage, AK. 252 pp.

Hodge, R.P. 1971. Herpetology north of 60°. The Beaver, Summer 1971:36—38.

Hodge, R.P. 1976. Amphibians and Reptiles in Alaska, The Yukon and Northwest Territories. Alaska Northwest Publishing, Anchorage, AK. xi + 89 pp. [reprinted 1977].

Logier, E.B.S. and G.C. Toner. 1955. **Check List of the Amphibians and Reptiles of Canada and Alaska**. Royal Ontario Museum Life Sciences Division Contribution 41. v + 88 pp.

Logier, E.B.S. and G.C. Toner. 1961. Check List of the Amphibians and Reptiles of Canada and Alaska. Royal Ontario Museum Life Sciences Division Contribution 53. vi + 92 pp.

MacDonald, S.O. 2003. **The Amphibians and Reptiles of Alaska**. Privately published, Juneau, AK. ii + 44 pp.

Arizona

Aitchison, S.W. and Tomko, D.S. 1974. Amphibians and reptiles of Flagstaff, Arizona. Plateau 47(1):18—25.

Brennan, T.C. and R.D. Babb. 2015. Herpetofauna of Arizona / Herpetofauna de Arizona. Pp. 144–163, 360–380 *in* Lemos-Espinal, J.A. (ed.), Amphibians and Reptiles of the US-Mexico Border States / Anfibios y Reptiles de los Estados de la Fontera México–Estados Unidos. Texas A&M University Press, College Station, TX. [checklist, plates and index collective for entire book].

Brennan, T.C. and A.T. Holycross. 2005. Amphibians and Reptiles of Maricopa County. Arizona Game and Fish Department, Phoenix, AZ. 68 pp.

Brennan, T.C. and A.T. Holycross. 2006. Amphibians and Reptiles in Arizona. Arizona

Game and Fish Dept., Phoenix, AZ. v + 150 pp.

Burns, G.T. n.d. Snakes and Other Venomous Creatures of Northwest Arizona. Hualapai Mountain Medical Center, Kingsman, AZ. 18 pp.

Cowles, R.B. and C.M. Bogert. 1936. The herpetology of the Boulder Dam Region. Herpetologica 1:33–42.

Coues, E. 1875. Synopsis of the reptiles and amphibians of Arizona, with critical and field notes, and an extensive synonymy. pp. 585–633, pls. 16–25 *in* G.M. Wheeler, Report upon Geographical and Geological Explorations and Surveys West of the One Hundredth Meridian, Vol. V, Zoology. Engineer Department, U.S. Army, Washington D.C. 1021 pp. [reprinted in 1978 *in* Adler, K. (ed.), Herpetological Explorations of the Great American West, Volume 1. Arno Press, New York, NY].

Dodge, N.N. 1938. Amphibians and reptiles of Grand Canyon National Park. Grand Canyon Natural History Association Bulletin 9:1–55.

Eaton, T.H., Jr. 1935. Report on amphibians and reptiles of the Navajo Country. Rainbow Bridge Monument Valley Expedition Bulletin 3:1–20.

Fowlie, J. 1965. The Snakes of Arizona. Azul Quinta Press, Fallbrook, CA. lv + 164 pp.

Grater, R.K. 1981. Snakes, Lizards, and Turtles of the Lake Mead Region. Southwest Parks and Monuments Association, Globe, AZ. 49 pp.

Hallowell, E. 1853. Reptiles. pp. 106–147, pls. 1–20 in L. Sitgreaves, Report of an Expedition down the Zuni and Colorado Rivers. US Senate, Executive [Document] No. 59. Washington, D.C. [reprinted 1962, The Rio Grande Press, Chicago, IL and 1978 *in* Adler, K. (ed.), Herpetological Explorations of the Great American West, Volume 2. Arno Press, New York, NY].

Holycross, A.T., T.C. Brennan, and R.E. Babb. 2022. A Field Guide to Amphibians and Reptiles of Arizona, Second Edition. Arizona Game and Fish Dept. Phoenix, AZ. 165 pp.

Holycross, A.T. and J.C. Mitchell (eds.). 2020. **Snakes of Arizona**. ECO Publishing, Rodeo, NM. 836 pp.

Howland, J.M., R. Babb, T.B. Johnson, S.G. Seim, M.J. Sredl, B.D. Taubert, and D.L. Waters. 1993. **Herps of Arizona**. Arizona Wildlife Views 36(3): 1–36. [reprinted in 1999 in a special edition of Arizona Wildlife Views, pp. 100–136].

Jones, K.B. 1981. Distribution, Ecology and Habitat Management of the Reptiles and Amphibians of the Hualapai–Aquarius Planning Area, Mohave and Yavapai Counties, Arizona. U.S. Department of the Interior, Bureau of Land Management Technical Note 353: vi + 134 pp.

Killian, J.L. 1954. Common Reptiles of Arizona. Arizona Game and Fish Department, Phoenix, AZ. 16 pp.

Lazaroff, D.W., P.C. Rosen, and C.H. Lowe, Jr. 2006. Amphibians, Reptiles, and Their Habitats at Sabino Canyon. University of Arizona Press, Tucson, AZ. xvi + 158 pp.

Lowe, C.H. 1955. The salamanders of Arizona. Transactions of the Kansas Academy of Sciences 58:237–251.

Lowe, C.H. 1964. Amphibians and Reptiles of Arizona pp. 153–174 in Lowe, C.H. (ed.), Vertebrates of Arizona. University of Arizona Press, Tucson, AZ. vii + 259 pp. [reprinted 1964, 1967, 1972, 1980, 2016; 2016 edition x + 270 pp.].

Lowe, C.H., C.R. Schwalbe, and T.B. Johnson. 1986. The Venomous Reptiles of Arizona. Arizona Game and Fish Department, Phoenix, AZ. ix + 115 pp.

McKee, E.D. and C.M. Bogert. 1934. The amphibians and reptiles of Grand Canyon National Park. Copeia 1934:178–180.

Miller, D.M., R.A. Young, T.W. Gatlin, and J.A. Richardson 1982. Amphibians and Reptiles of the Grand Canyon. Grand Canyon Natural History Association Monograph 4. vii + 144. pp

Murphy, J.C. 2018. Arizona's Amphibians and Reptiles: A Natural History and Field Guide. Book Services (www.bookservices.org). xii + 316 pp.

Murphy, J.C. 2019. Arizona's Amphibians and Reptiles: A Natural History and Field Guide. JCM Natural History Publishing, Green Valley, AZ. 348 pp. [revised and expanded version of Murphy 2018].

Nelson, D. and S. Nelson. 1977. Easy Field Guide to Common Snakes of Arizona. Tecolote Press, Glenwood, NM. 32 pp. (unnumbered). [reprinted in 1985 by Primer Publishers, Phoenix, AZ and 1994 by American Traveler Press, Phoenix, AZ].

Rosen, P.C., and C.H. Lowe. 1996. Ecology of the Amphibians and Reptiles at Organ Pipe Cactus National Monument, Arizona. United States Department of the Interior, National Biological Service, Cooperative Park Studies Unit, The University of Arizona, and National Park Service, Organ Pipe Cactus National Monument, Technical Report No. 53. 136 pp.

Ruthven, A.G. 1907. A collection of reptiles and amphibians from southern New Mexico and Arizona. Bulletin of the American Museum of Natural History 23:483–604.

Schuett, G.W., C.F. Smith, and B. Ashley. 2018. Rattlesnakes of the Grand Canyon. ECO Publishing, Rodeo, NM. 131 pp.

Schuett, G.W., C.F. Smith, and W. Wells. 2023. Amphibians of the Sky Islands. ECO Publishing, Rodeo, NM. 144 pp.

Schuett, G.W., M.J. Feldner, C.F. Smith, and R.S. Reiserer (eds.). 2016. Rattlesnakes of Arizona. ECO Publishing, Rodeo, NM. 2 Volumes 739 pp. + 488 pp.

Smith, R.L. 1982. Venomous Animals of Arizona. Cooperative Extension Service, College of Agriculture, University of Arizona Bulletin 8245. xvi + 134 pp. [amphibians and reptiles on pp. 97–114].

Stejneger, L. 1902. The reptiles of the Huachuca Mountains, Arizona. Proceedings of the United States National Museum 1282:149–158.

Tomko, D.S. 1975. The reptiles and amphibians of the Grand Canyon. Plateau 47:161–166.

Turner, D.S., P.A. Holm, E.B. Wirt, and C.R. Schwalbe. 2003. Amphibians and reptiles of the Whetstone Mountains, Arizona. Southwestern Naturalist 48(3):347–355.

Van Denburgh, J., and J.R. Slevin. 1913. A list of the amphibians and reptiles of Arizona, with notes on the species in the collection of the Academy. Proceedings of the California Academy of Sciences 3:391–454, pls.17–28.

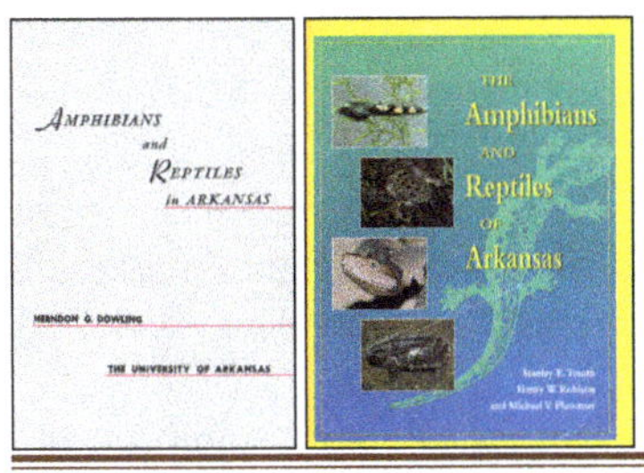

Arkansas

Black, J.D. and S.C. Dellinger. 1938. Herpetology of Arkansas, Part II: The amphibians. Occasional Papers of the University of Arkansas Museum 2:1–30.

Dellinger, S.C. and J.D. Black. 1938. Herpetology of Arkansas, Part I: The reptiles. Occasional Papers of the University of Arkansas Museum 1:1–47.

Dowling, H.G. 1957. **A review of the amphibians and reptiles of Arkansas**. Occasional Papers of the University of Arkansas Museum 3. 51 pp.

Hurter, J. and J.K. Strecker, Jr. 1909. Amphibians and reptiles of Arkansas. Transactions of the Academy of Sciences of St. Louis 18:11–27.

Irwin, K. 2004. Arkansas Snake Guide. Arkansas Game and Fish Commission, Little Rock, AR. 50 pp.

Plummer, M.V. 1985. Arkansas turtles. Arkansas Game and Fish 16(3):13–20.

Reagen, D.P. 1974. Threatened native amphibians in Arkansas. pp. 93–99 *in* Arkansas Natural Area Plan. Arkansas Department of Planning, Little Rock, AR. 243 pp.

Reagen, D.P. 1974. Threatened native reptiles in Arkansas. pp. 101–105 *in* Arkansas Natural Area Plan. Arkansas Department of Planning, Little Rock, AR. 243 pp.

Schwardt, H.H. 1938. Reptiles of Arkansas. University of Arkansas Agricultural Experiment Station Bulletin 357:1– 47.

Sutton, K. 1987. Arkansas' venomous snakes. Arkansas Game and Fish 18(3):13–20.

Trauth, S. 1975. Lizards of Arkansas. Ozark Society Bulletin 1974–1975:1–4.

Trauth, S.E., H.W. Robison, and M.V. Plummer. 2004. **The Amphibians and Reptiles of Arkansas**. University of Arkansas Press, Fayetteville, AR. xvii + 421 pp.

Vance, T. 1982. A field key to the reptiles of Arkansas. Bulletin of New York Herpetological Society 17:32– 48.

Vance, T.L. 1985. Annotated checklist and bibliography of Arkansas reptiles. Smithsonian Herpetological Information Service 63. 45 pp.

California

Atsatt, S.R. 1913. The reptiles of the San Jacinto area of southern California. University of California Publications in Zoology 12:31–50.

Badaracco, R. 1962. Amphibians and Reptiles of Lassen Volcanic National Park. National

Park Service, Mineral, CA. 59 pp.

Banta, B.H. 1962. A Preliminary account of the herpetofauna of the Saline Valley Hydrographic Basin, Inyo County, California. Wasmann Journal of Biology 20:161–251.

Banta, B.H. and D. Morafka. 1966. An annotated checklist of the recent amphibians and reptiles inhabiting the city and county of San Francisco, California. Wasmann Journal of Biology 24: 223–238.

Banta, B.H. and D. Morafka. 1968. An annotated check list of the Recent amphibians and reptiles of the Pinnacles National Monument and Bear Valley, San Benito and Monterey Counties, California with some ecological observations. Wasmann Journal of Biology 26:161–183.

Basey, H.E. (ed.). 1962. Amphibians of Tulare County. Tulare County Superintendent of Schools Office, Visalia, CA. ii + 16 pp.

Basey, H.E. 1969. Sierra Nevada Amphibians. Sequoia Natural History Association, Three Rivers, CA. 27 pp.

Basey, H.E. 1976. Discovering Sierra Reptiles and Amphibians. Yosemite and Sequoia Natural History Associations, Three Rivers, CA. vi + 50 pp. [reprinted 1988, 1991].

Bogert, C.M. 1930. An annotated list of the amphibians and reptiles of Los Angeles County, Calif. Bulletin of the Southern California Academy of Sciences 29:1–14.

Brown, P.R. 1997. A Field Guide to Snakes of California. Gulf Publishing Co., Houston, TX vii + 215 pp.

Camp, C.L. 1916. Notes on the local distribution and habits of the amphibians and reptiles of southeastern California in the vicinity of the Turtle Mountains. University of California Publications in Zoology 12:503–544.

Cornett, J.W. 1996. Rattlesnakes of the California Deserts. Palm Springs Desert Museum, Palm Springs, California. [iv] + 13.

Cronise, T.F. 1868. The Natural Wealth of California. H. H. Bancroft and Company, San Francisco. xvi + 696 pp. [amphibians and reptiles on pp. 480–486].

DeLisle, H.F., Gilbert C., J. Feldner, P. O'Connor, M. Peterson, and P. Brown. 1986. The Distribution and Present Status of the Herpetofuana of the Santa Monica Mountains of Los Angeles and Ventura Counties, California. Southwest Herpetological Society Special Publication No. 2. 94 pp.

DeLisle, H.F., R.H. Horne, and P. O'Connor. 1985. An annotated checklist of the herpetofauna of the Santa Monica Mountains, California. Herpton 9:1–17.

Dixon, J.R. 1967. Amphibians and Reptiles of Los Angeles County, California. Los Angeles Museum of Natural History, Science Series 23, Zoology 10. 64 pp.

Doudnik, N. 1955. Amphibians of Central and Northern California. Science Publication 9, Sacramento State College. [ii + 31 pp] [dated May 28, 1953, copyrighted 1955].

Fisher, R.N., and T.J. Case. 1997. A Field Guide to the Reptiles and Amphibians of Coastal Southern California. Privately published, La Jolla, California. 46 pp.

Flaxington, W.C. 1998. Field Observations of California Amphibians and Reptiles. Research Guide to the Activity and Distribution of California's Herpetofauna. Privately Published, Whittier, CA. vi + 129 pp.

Flaxington, W.C. 2021. Amphibians and Reptiles of California. Fieldnotes Press, Anaheim, CA. iv + 294 pp.

Glaser, H.S.R. 1970. The Distribution of Amphibians and Reptiles in Riverside County, California. Riverside Museum Press, Natural History Series No. 1. 40 pp.

Grinnell, J. and C.L. Camp. 1917. A distributional list of the amphibians and reptiles of California. University of California Publications in Zoology 17:127–208.

Grinnell, J. and H.W. Grinnell. 1907. Reptiles of Los Angeles County, California. Bulletin of the Throop Institute 35:1–64.

Grinnell, J. and T.I. Storer. 1921. Reptiles and amphibians of Yosemite National Park. pp. 175–182 *in* A.F. Hall, (ed.), Handbook of Yosemite National Park. G. P. Putnam's Sons, New York and London.

Grinnell, J. and T.I. Storer. 1924. Animal Life of the Yosemite. University of California Press, Berkeley, CA. xviii + 752 pp. [amphibians and reptiles on pp. 626–666].

Hill, H.R. 1948. Amphibians and Reptiles of Los Angeles County. Los Angeles County Museum, Science Series No. 12, Zoology 5. 30 pp.

Hirsch, R., S. Hathaway, and R. Fisher. 2002. Herpetofauna and Small Mammal Surveys on the Marine Corps Air Ground Combat Center, Twentynine Palms, CA. US Geological Survey, Sacramento, CA. iv + 20 pp.

Hollingsworth, B.D. and C.R. Mahrdt. 2015. Herpetofauna of California / Herpetofauna de California. Pp. 122–143, 336–359 *in* Lemos-Espinal, J.A. (ed.), Amphibians and Reptiles of the US-Mexico Border States / Anfibios y Reptiles de los Estados de la Fontera México–Estados Unidos. Texas A&M University Press, College Station, TX. [checklist, plates and index collective for entire book].

Jennings, M.R. 1983. Annotated checklist of the amphibians and reptiles of California. California Fish and Game 69:151–171.

Jennings, M.R. 1987. Annotated Checklist of the Amphibians and Reptiles of California, 2nd, Revised Edition. Southwestern Herpetologists Society Special Publication No. 3. 48 pp.

Jennings, M.R. 2004. An annotated checklist of the amphibians and reptiles of California and adjacent waters. California Fish and Game 90:161–213.

Jennings, M.R. and M.P. Hayes. 1994. Amphibians and Reptile Species of Special Concern in California. California Department of Fish and Game. Rancho Cordova, CA. iii + 255 pp.

Klauber, L.M. 1928. A list of the amphibians and reptiles of San Diego County, California. Bulletin of the Zoological Society of San Diego 4:1–8. [2nd edition 1930 as Bulletin of the Zoological Society of San Diego 5; originally privately distributed as a mimeographed list in 1926 [i] + 6 + [1] pp.].

Klauber, L.M. 1934. Annotated list of the amphibians and reptiles of the southern border of California. Bulletin of Zoological Society of San Diego 31:1–28.

Laudenslayer, W.F. and W.E. Grenfell, Jr. (eds.). 1983. A list of amphibians, reptiles, birds, and mammals of California. Outdoor California 44(1):5–14.

Laudenslayer, W.F., W.E. Grenfell, Jr, and D.C. Zeiner. 1991. A checklist of the amphibians, reptiles, birds, and mammals of California. California Fish and Game 77:109–141.

Lemm, J.M. 2006. Field Guide to Amphibians and Reptiles of the San Diego Region. University of California Press, Berkeley, CA. xii + 326 pp.

McKeown, S. 1974. Checklist of amphibians and reptiles of Santa Barbra County, California. Santa Barbra Museum of Natural History Occasional Paper 9. 11 pp.

Morafka, D. 1968. The Herpetofauna of City and County of San Francisco, California. Student Discovery, Occasional Papers of the Junior Academy, California Academy of Sciences, San Francisco. 20 + [2] pp.

Parker, J.M. and S. Brito. 2013. Reptiles and Amphibians of the Mojave Desert. A Field Guide. Snell Press, Las Vegas, NV. 184 pp.

Perkins, C.B. 1938. The snakes of San Diego County with descriptions and key. Bulletin of Zoological Society of San Diego 13:1–66.

Perkins, C.B. 1949. The snakes of San Diego County with descriptions and key, second ed. Bulletin of Zoological Society of San Diego 23:1–77.

Persons, T.B., and E.M. Nowak. 2007. Inventory of Amphibians and Reptiles at Mojave National Preserve. Final Report. U.S. Department of the Interior, U. S. Geological Survey Open–File Report 2007–1109. i + 72 pp.

Pickwell, G. 1947. Amphibians and Reptiles of the Pacific States. Stanford University Press, Stanford, CA. xiv + 236 pp. [reprinted 1949; reprinted 1972 by Dover Publications, New York, NY, xviii + 234 pp. with a new foreword and table of changes in nomenclature by G. V. Pickwell].

Savage, J.M. 1948. An Illustrated Key to the Lizards, Snakes and Turtles of California. Volume 1. Naturegraph Pocket Keys. Naturegraph Company, Los Altos, CA. 16 pp.

Schoenherr, A.A. 1976. The Herpetofauna of the San Gabriel Mountains. Southwestern Herpetologists Society, Los Angeles, CA. 95 pp.

Shaw, C.E. 1950. The lizards of San Diego County with descriptions and key. Bulletin of the Zoological Society of San Diego 25:1–63.

Shedd, J.D. 2005. Amphibians and Reptiles of Bidwell Park. Quadro Publishing, Chico, CA. 118 pp.

Slevin, J.R. 1934. A Handbook of Reptiles and Amphibians of the Pacific States. California Academy of Sciences, San Francisco, CA. 73 pp.

Sloan, A.J. 1964. Amphibians of San Diego County. Occasional Papers of the San Diego Society of Natural History 13. 42 pp.

Stebbins, R.C. 1959. Reptiles and Amphibians of the San Francisco Bay Region. University of California Press, Berkeley, CA. 73 pp. (Natural History Guide 3)

Stebbins, R.C. 1972. Reptiles and Amphibians of California. University of California Press., Berkeley, CA. 152 pp. (Natural History Guide 31).

Stebbins, R.C. and S.M. McGinnis. 2012. **Field Guide to Amphibians and Reptiles of California**, Revised Edition. University of California Press, Berkeley, CA. xiii + 538 pp.

Stejneger, L. 1893. Annotated List of the Reptiles and Batrachians Collected by the Death Valley Expedition in 1891, with Descriptions of New Species. pp. 159–228 *in* The Death Valley Expedition Part II. North American Fauna No. 7. 402 pp.

Stephens, F. 1921. An annotated list of the amphibians and reptiles of San Diego County, California. Transactions of San Diego Society of Natural History 3:57–69.

Storer, T.I. 1925. A synopsis of the Amphibia of California. University of California Publications in Zoology 27:1–342.

Thomson, R.C., A.N. Wright, and H.B. Shaffer. 2016. California Amphibian and Reptile Species of Special Concern. University of California Press, Oakland, CA. 390 pp.

Turner, F.B. and R.H. Wauer. 1963. A Survey of the herpetofauna of the Death Valley area. Great Basin Naturalist 23:119–128.

Wade, D. and J. Hooper. 1980. California Rattlesnakes. Univ of California - Div. of Agricultural Sciences. Berkeley, Leaflet 2996. 7 pp.

Walker, M.V. 1946. Reptiles and amphibians of Yosemite National Park. Yosemite Nature Notes 25(1): 1–48. [reprinted 1957 and in an unknown number of subsequent reprints without date and without indication of "Yosemite Nature Notes"].

Zeiner, D.C., W.F. Laudenslayer, Jr. and K.E. Mayer. 1988. **California's Wildlife Volume 1: Amphibians and Reptiles**. California Department of Fish and Game, Sacramento, CA. ix + 272 pp.

Colorado

Barry, L.T. 1933. Snakes of the Mesa Verde National Park. Mesa Verde Notes 4(2):8–11.

Carl, G., C. Counts, M. Hosrt, and L. Livo. 1976. A Guide to the Reptiles and Amphibians of the Denver Area, Including the Counties of Jefferson, Arapahoe, Boulder, Denver and Adams. Colorado Herpetological Society, Denver, CO. 20 pp.

Cockerell, T.D.A. 1927. Zoology of Colorado. University of Colorado, Boulder, CO. vii + 262 pp., 6 pls. [amphibians and reptiles on pages 104—113].

Douglas, C.L. 1966. Amphibians and reptiles of Mesa Verde National Park, Colorado. University of Kansas Publications of the Museum of Natural History 15 :711–744.

Ellis, M.M. and J. Henderson. 1913. The amphibians and reptiles of Colorado, Part 1. University of Colorado Studies 10:39–129 + 8 pls.

Ellis, M.M. and J. Henderson. 1915. The amphibians and reptiles of Colorado, Part 2. University of Colorado Studies 15:253–263.

Garber, S.D. 1977. A preliminary review of the Rocky Mountain National Park herpetofauna. Kansas Herpetological Society Newsletter. 21:13–16.

Hammerson, G.A. 1982. Amphibians and Reptiles in Colorado. Colorado Division of Wildlife, Denver, CO. vii + 131 pp.

Hammerson, G.A. 1999. **Amphibians and Reptiles in Colorado, Revised Edition**. University Press of Colorado. Boulder, CO. xxvi + (2) + 484pp.

Hammerson, G.A. and D. Langlois (eds.). 1981. Colorado Reptile and Amphibian Distribution Latilong Study, 2nd Edition. Colorado Division of Wildlife, Denver, CO. 24 pp.

Jones–Burdick, W.H. 1939. Guide to the Snakes of Colorado. University of Colorado Museum Leaflet 1. 12 pp.

Jones–Burdick, W.H. 1949. Guide to the Snakes of Colorado (Revised). University of Colorado Museum Leaflet 5. 24 pp. [reprinted 1963].

Langlois, D. 1978. Colorado Reptile and Amphibian Distribution Latilong Study. Colorado Division of Wildlife, Denver, CO. [ii] + 17 + [2] pp.

Livo, L.J. 1985. The small snakes. Colorado Outdoors 34(4):36–39.

Livo, L.J. 1995. Colorado Amphibian and Reptile Records by County. Colorado Herpetological Society, Denver, CO. 20 pp.

Livo, L.J. 1995. Identification Guide to Montane Amphibians of Southern Rocky Mountains. Colorado Division of Wildlife, Denver, CO. 25 pp.

Maslin, T.P. 1947. Guide to the Lizards of Colorado. University of Colorado Museum Leaflet 3. 14 pp.

Maslin, T.P. 1959. An Annotated Checklist of the Amphibians and Reptiles of Colorado. University of Colorado Studies, Series in Biology 6. 98 pp.

Maslin, T.P. 1964. Amphibians and reptiles of the Boulder area. Pp. 75–80 in Rodeck, H. G. (ed.), The Natural History of the Boulder Area. University of Colorado Museum Leaflet 13. iii + 100 pp.

Rodeck, H.G. 1943. Guide to the Amphibia of Colorado. University of Colorado Museum Leaflet 2. 8 pp. [reprinted in 1950].

Rodeck, H.G. 1950. Guide to the Turtles of Colorado. University of Colorado Museum Leaflet 7. 9 pp. + 6 pls.

Smith, H.M., T. P. Maslin, and R. L. Brown. 1965. Summary of the Distribution of the Herpetofauna of Colorado. University of Colorado Studies, Series in Biology 15. 52 pp.

Young, M.T. 2011. **Guide to Colorado Reptiles and Amphibians**. Fulcrum Press, Golden, CO. vi + 170 pp.

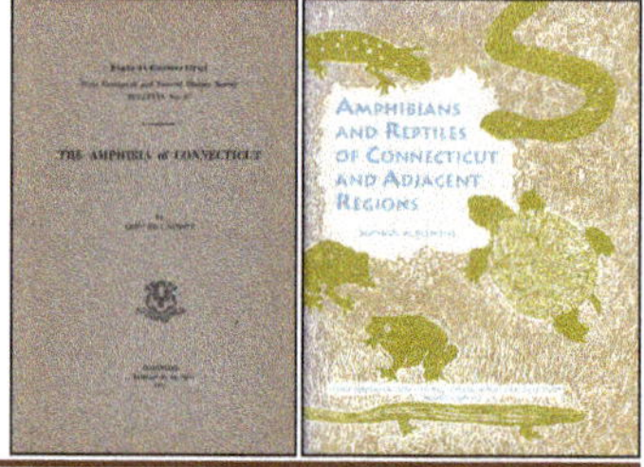

Connecticut

Babbitt, L.H. 1937. **The Amphibia of Connecticut**. Connecticut Geological and Natural History Survey Bulletin 57. 50 pp. + 20 pls.

Devito, J. and J. Markow. 1998. Amphibians and Reptiles of the Connecticut College Arboretum. Connecticut College, New London, CT. 52 pp.

Dickson, J. and J. Victoria. 1997. Snakes in Connecticut. Connecticut Department of Environmental Protection. 18 pp.

Klemens, M.W. 1991. Checklist of the Amphibians and Reptiles of Connecticut with Notes on Uncommon Species. Connecticut Department of Environmental Protection Bulletin 14. 24 pp.

Klemens, M.W. 1993. **Amphibians and Reptiles of Connecticut and Adjacent Regions**. Connecticut Geological and Natural History Survey Bulletin 112. 318 pp.

Klemens. M.W. 2000. Amphibians and Reptiles in Connecticut: A Checklist. Connecticut Department of Environmental Protection Bulletin 32. 96 pp.

Klemens, M.W., H.J. Gruner, D.P. Quinn, and E.R. Davison. 2021. Conservation of Amphibians and Reptiles in Connecticut. Connecticut Department of Energy and Environmental Protection, Hartford, CT. 305 pp.

Lamson, G.H. 1935. The Reptiles of Connecticut. Connecticut Geological and Natural History Survey Bulletin 54. 35 pp. + 12 pls.

Linsley, J.H. 1844. A catalogue of the reptiles of Connecticut arranged according to their families. American Journal of Science and Arts 46:37–51.

Watkins–Colwell, G.J., T.A. Leenders, B.J. Roach, D.J. Drew, G. Dancho, and J. Yuckienuz. 2006. New distribution records for the amphibians and reptiles in Connecticut, with notes on the status of an introduced species. Bulletin of the Peabody Museum Natural History 47:47–62.

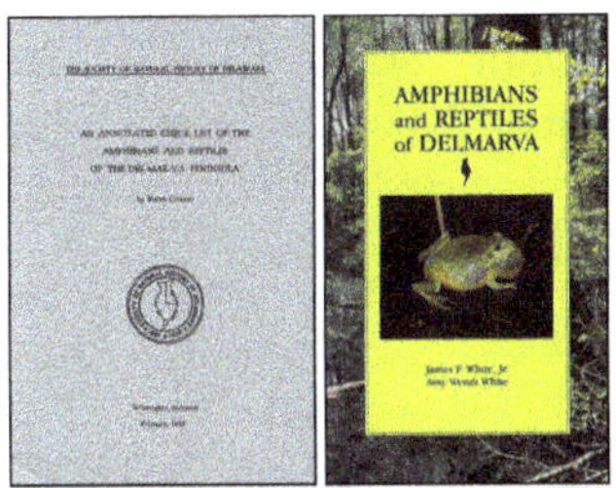

Delaware

Arndt, R.G. 1975. Turtles of Delaware/Meet our turtles. Delaware Conservationist 19(1):10–15.

Arndt, R.G. 1976. Delaware snakes. Delaware Conservationist 20(1):11–14.

Conant, R. 1945. **An Annotated Checklist of the Amphibians and Reptiles of the Del-Mar-Va Peninsula**. Society of Natural History of Delaware, Wilmington, DE. 8 pp.

Conant, R. 1947. Reptiles and amphibians in Delaware. pp. 23–25 *in* Delaware, a History of the First State. H.C. Reed, ed. Lewis Historical Publishing Co., New York, NY.

Hardy, Jr., J.D. 1972. Amphibians of the Chesapeake Bay region. Chesapeake Science 13 (supplement):123–128.

Hardy, Jr., J.D. 1972. Reptiles of the Chesapeake Bay region. Chesapeake Science 13 (supplement):128–134.

Reed, C.F. 1956. Contributions to the Herpetology of Maryland and Delmarva, 5. Bibliography to the Herpetology of Maryland, Delmarva, and the District of Columbia. Reed Herpetorium (privately published), Baltimore, MD. 9 pp.

Reed, C.F. 1956. Contributions to the Herpetology of Maryland and Delmarva, 6. An Annotated Check List of the Lizards of Maryland and Delmarva. Reed Herpetorium (privately published), Baltimore, MD. 6 pp.

Reed, C.F. 1956. Contributions to the Herpetology of Maryland and Delmarva, 7. An Annotated Check List of the Turtles of Maryland and Delmarva. Reed Herpetorium (privately published), Baltimore, MD. 11 pp.

Reed, C.F. 1956. Contributions to the Herpetology of Maryland and Delmarva, 8. An Annotated Check List of the Snakes of Maryland and Delmarva. Reed Herpetorium (privately published), Baltimore, MD. 20 pp.

Reed, C.F. 1956. Contributions to the Herpetology of Maryland and Delmarva, 9. An Annotated Check List of the Frogs and Toads of Maryland and Delmarva. Reed Herpetorium (privately published), Baltimore, MD. 19 pp.

Reed, C.F. 1956. Contributions to the Herpetology of Maryland and Delmarva, 11. An Annotated Herpetofauna of the Delmarva Peninsula, including Many New or Additional Localities. Reed Herpetorium (privately published), Baltimore, MD. 11 pp. + 1 p. unnumbered.

Reed, C.F. 1957. Contributions to the Herpetology of Maryland and Delmarva, 10. An Annotated Check List of the Salamanders of Maryland and Delmarva. Reed Herpetorium (privately published), Baltimore, MD. 6 pp.

Reed, C.F. 1957. Contributions to the herpetology of Maryland and Delmarva, 15: The herpetofauna of Somerset County, Md. Journal of the Washington Academy of Sciences 47:127–128.

White, Jr., J.F. and A.W. White. 2002. **Amphibians and Reptiles of Delmarva.** Tidewater Publishers, Centreville, MD. xvi + 248 pp., 32 pp. pls.

White, Jr., J.F. and A. W. White. 2007. Amphibians and Reptiles of Delmarva. Second Edition. Delaware Nature Society, Inc., Tidewater Publishers, Centreville, Maryland. xvi + 243 pp., 32 pp. pls. [accompanied by an addendum sheet dated 2008, reprinted 2009].

Florida

Allen, R. 1940. Poisonous Snakes of Florida. Florida Fish and Game Magazine 1(10):6–9.

Allen, R. and W.T. Neill 1949. A Checklist of the Amphibians and Reptiles of Florida. Ross Allen's Reptile Institute. Silver Springs, FL. 4 pp.

Allen, R. 1963. Florida's snakes and lizards. pp. 54–59 *in* Maxwell, L.S. 1963. Florida's Poisonous Plants, Snakes, Insects. Lewis S. Maxell Publisher, Tampa FL. 72 pp.

Anderson, R. 1975. Guide to Florida Poisonous Snakes. Central Press, Hialeah, FL. [ii] + 53 pp. [revised 1983 by Anderson's Nature Guides, Port Richey, FL, [ii] + 59 pp., and 1985 by Winner Enterprises, Altamonte Springs, FL, iv + 56 pp.].

Anderson, R. 1985. Guide to Florida Alligator and Crocodile. Winner Enterprises, Altamonte Springs, FL. iv + 56 pp.

Anderson, R. 1985. Guide to Florida Nonpoisonous Snakes. Winner Enterprises, Altamonte Springs, FL. iv + 56 pp.

Anderson, R. 1985. Guide to Florida Poisonous Snakes. Winner Enterprises, Altamonte Springs, FL. iv + 56 pp.

Anderson, R. 1985. Guide to Florida Turtles – Sea Turtles included. Winner Enterprises, Altamonte Springs, FL. iv + 56 pp.

Anderson, R. 1988. Guide to Florida Lizards and Amphibians. Winner Enterprises, Altamonte Springs, FL. iv + 56 pp.

Anderson, R. 1989. The Great Outdoors Book of Florida Snakes. Great Outdoors Publishing Co., St. Petersburg, FL. 127 pp.

Arnold, D. 1995. Florida's Marine Turtles. Florida Wildlife. 49(4):12–15.

Ashton, Jr., R.E. 1977. Identification Manual to the Reptiles and Amphibians in Florida. Florida State Museum, Public Series No. 1. 65 pp.

Ashton, Jr., R.E. and P.S. Ashton. 1981. Handbook of Reptiles and Amphibians of Florida, Part One: The Snakes. Windward Publishing, Miami, FL. 176 pp.

[revised edition 1988].

Ashton, Jr., R.E. and P.S. Ashton. 1985. Handbook of Reptiles and Amphibians of Florida, Part Two: Lizards, Turtles, and Crocodilians. Windward Publishing, Miami, FL. 191 pp. [2nd edition 1991].

Ashton, Jr., R.E. and P.S. Ashton. 1988. Handbook of Reptiles and Amphibians of Florida, Part Three: The Amphibians. Windward Publishing, Miami, FL. 191 pp.

Bartlett, R.D. and P.P. Bartlett. 1999. A Field Guide to Florida Reptiles and Amphibians. Gulf Publishing Company, Houston, TX. xvi + 280 pp.

Bartlett, R.D. and P.P. Bartlett. 2004. Florida Snakes. University Press of Florida, Tallahassee, FL. xiv + 182 pp.

Bartlett, R.D. and P.P. Bartlett. 2011. Florida's Frogs, Toads, and Other Amphibians. University of Florida Press, Gainesville, FL. 188 pp.

Bartlett, R.D. and P.P. Bartlett. 2011. Florida's Turtles, Lizards, and Crocodilians University of Florida Press, Gainesville, FL. 257 pp.

Bartolotti, K. and G. Bartolotti. 2010. A Field Guide to Snakes of Hernando, Hillsborough, Manatee, Pasco, and Pinellas Counties in Florida. Robertson Publishing, Los Gatos, CA. 39 pp.

Carmichael, P. and W. Williams. 1991. Florida's Fabulous Amphibians and Reptiles. World Publishing, Tampa, FL. 121 pp.

Carr, A.F. 1940. A Contribution to the Herpetology of Florida. University of Florida Publication, Biological Sciences Series 3. 118 pp.

Carr. A.F. and C.J. Goin. 1955. **Guide to the Reptiles, Amphibians and Fresh-Water Fishes of Florida.** University of Florida Press, Gainesville, FL. xxviii + 341 pp. [reissued in hardcover 1959, 1969].

Collins, J.T., S.L. Collins, and T.W. Taggart. 2011. A Pocket Guide to the Snakes of St. Vincent National Wildlife Refuge– Florida. Center for North American Herpetology, Lawrence, KS. 48 pp.

Duellman, W.E. and A. Schwartz. 1958. Amphibians and reptiles of southern Florida. Bulletin of Florida State Museum 3:181–324.

Enge, K.M. 2002. An Updated, Indexed Bibliography of the Herpetofauna of Florida. Florida Fish and Wildlife Conservation, Tallahassee, FL. iii + 410 pp.

Enge, K.M. n.d. [2008]. Venomous and Nonvenomous Snakes of Florida. Florida Fish and Wildlife Conservation, Tallahassee, FL. 16 pp.

Enge, K.M. and C.K. Dodd, Jr. 1986. A bibliography of the herpetofauna of Florida. Smithsonian Herpetological Information Service 72. 68 pp.

Enge, K.M. and C.K. Dodd, Jr. 1992. An Indexed Bibliography of the Herpetofauna of Florida. Florida Game and Fresh Water Fish Commission Nongame Wildlife Program Technical Report No. 11. v + 231 pp.

Enge, K.M. and K.N. Wood. 2001. Herpetofauna of Chinsegut Nature Center, Hernando County, Florida. Florida Scientist 64:283–305.

Godwin, O. 1961. Snakes of Florida. Gatorland, Kissimmee, FL. 48 pp. [revised in 1973 and 1981; at least five printings in total; publisher variously listed as Gatorland or Gatorland Zoo; 5th printing (1981) published in Orlando].

Haast, W.E. and R. Anderson. 1981. Complete Guide to Snakes of Florida. The Phoenix Publishing Company, Inc., Miami, FL. 139 pp.

Holbrook, J.D. 2012. A Field Guide to the Snakes of Southern Florida. ECO Publishing, Rodeo, NM. 179 pp.

Hollister, J.M. 1951. Turtles found in Florida. Florida Naturalist 24:93–95.

Iverson, J.B. and C.R. Etchberger. 1989. Distribution of the turtles of Florida. Florida Scientist 52:119–144.

Jensen, A.S. 1974. A Checklist of Native Florida Snakes. Florida Cooperative Extension Service Wildlife Report 75(2):1–4.

Jensen, A.S. 1981. Poisonous Snakes of Florida. Florida Cooperative Extension Service Forest Resources and Conservation Fact Sheet 9. 3 pp.

Koukoulis, A. 1972. Poisonous Snakes of Florida. International Graphics, Inc. Hollywood, FL. 32 pp

Krysko, K.L., K. Enge, and P.E. Moler. 2011. Atlas of Amphibians and Reptiles in Florida. Final Report, Project Agreement Number 08013, Florida Fish and Wildlife Conservation Commission, Tallahassee, Florida. iv + 524 pp.

Krysko, K.L., K. Enge, and P.E. Moler. 2019. **Amphibians and Reptiles of Florida**. University of Florida Press, Gainesville, FL. 706 pp.

Lazell, Jr., J. D. 1989. Wildlife of the Florida Keys, A Natural History. Island Press, Washington, D.C. xvi + 254 pp. [amphibians and reptiles on pages 109–216].

LeBuff, C. and C. Lechowicz. 2014. Amphibians and Reptiles of Sanibel and Captiva Islands, Florida. Amber Publishing, Fort Myers, FL. 279 pp.

Lee, D.S. 1970. A list of the amphibians and reptiles of Florida. Bulletin of Maryland Herpetological Society 6:74–80.

Lockey, R.F. and L.S. Maxwell. 1978. Florida's Poisonous Plants, Snakes, Insects. Lewis S. Maxwell Publisher, Tampa, FL. 79 pp. [revised 1986 and 1992].

McDiarmid, R.W. (ed). 1978. Rare and Endangered Biota of Florida: Volume Three: Amphibians and Reptiles. University Press of Florida, Gainesville, FL. xxii + 74 pp.

Means, D.B. 1999. Venomous snakes of Florida. Florida Wildlife 53(5):13–20.

Meshaka, Jr., W.E. 2011. A Runaway Train in the Making: The Exotic Amphibians, Reptiles, Turtles, and Crocodilians of Florida. Monograph 1. Herpetological Conservation and Biology. 6:1–101.

Meshaka, Jr., W.E. and K.J. Babbitt. 2005. Amphibians and Reptiles: Status and Conservation in Florida. Krieger Publishing, Malabar, FL. xvi + 317 pp.

Meshaka, Jr., W.E., B.P. Butterfield, and J.B. Hauge. 2004. The Exotic Amphibians and Reptiles of Florida. Krieger Publishing, Malabar, FL. x + 155 pp.

Meshaka, Jr., W.E. and J.N. Layne. 2015. The Herpetology of Southern Florida. Monograph 5. Herpetological Conservation and Biology. 10:1–353.

Meylan, P.A. 2006. Biology and Conservation of Florida Turtles. Chelonian Research Foundation, Lunenberg, MA. 376 pp.

Moler, P.E. 1988. A Checklist of Florida's Amphibians and Reptiles. Florida Game and Freshwater Fish Commission, Tallahassee, FL. pp. 3–18 (no pp. 1–2) [revised editions in 1990, 1999 (possibly more); 1999 edition with iv + 14 pp.].

Moler, P.E. 1991. Salamanders of Florida. Florida Wildlife 45(1):19–21

Moler, P.E. 1992. Rare and Endangered Biota of Florida, Vol III. Amphibians and Reptiles. University Press of Florida. Gainesville, FL. xxvii + 291 pp.

Neill, W.T. 1952. The reptiles of Florida: Part 1. Florida Naturalist 25:11–16.

Peebles, D. n.d. [2000]. Turtles of Pinellas County and its Surrounding Waters. Pinellas County, FL. 63 pp.

Pratt, K.K. 1993. Checklist of Florida Reptiles and Amphibians. F.E.W. Bradenton, FL. 65 pp.

Stevenson, H.M. 1976. Vertebrates of Florida: Identification and Distribution. University

Press of Florida, Gainesville, FL. xix + 607 pp. [amphibians and reptiles on pages 226– 290].

Tennant, A. 1997. A Field Guide to Snakes of Florida. Gulf Publishing Company, Houston, TX. xiii + 257 pp.

Tennant, A. 2003. Snakes of Florida, second edition. Lone Star Field Guide, Taylor Trade Publishing, Lanham, Maryland. xiii + [1] + 271 pp.

Truitt, J.O. 1962. A Guide to the Snakes of South Florida Lake Okeechobee to Florida Keys. Hurricane House Publishers, Miami, FL. v + 46 pp. [reprinted in 1966, 1968].

Truitt, J.O. and L.D. Ober. 1971. A Guide to the Lizards of South Florida. Hurricane House Publishers, Miami, FL. 37 pp.

Van Metter, V.B. 1983. Florida's Sea Turtles. Florida Power and Light Co., Miami, FL. [iii] + 46 pp. [reprinted 1986, 1987, 1988, 1990 ([iii] + 63 pp.), 1992, 2002 ([iii] + 60 pp.)].

Wilson, L.D. and L. Porras. 1983. The Ecological Impact of Man on the South Florida Herpetofauna. University of Kansas Museum of Natural History Special Publication No. 9. 89 pp.

Zappalorti, R. 2007. Florida's venomous snakes. Florida Wildlife 60(5):35–38.

Georgia

Camp, C.D., Jensen, J.B. and Elliott, M. n.d. [2004]. Salamanders of the Chattahoochee and Oconee National Forests. U.S. Department of Agriculture, Forest Service, Southern Region, n.p. 115 pp.

Gibbons, W. and P.J. West (eds.) 1998. Snakes of Georgia and South Carolina. Savannah River Ecology Laboratory HerpOutreach Publication 1. 28 pp.

Holbrook, J. 1849. Reptiles pp. 13–15 *in* [Appendix] Catalogue of the fauna and flora of the State of Georgia *in* White, G. (ed.), Statistics of the State of Georgia: Including an Account of its Natural, Civil, and Ecclesiastical History; Together with a Particular Description of each County, Notices of the Manners and Customs of its Aboriginal Tribes, and a Correct Map of the State. W. Thorne Williams, Savannah, GA. [includes amphibians; reprinted 1978 *in* Adler, K. (ed.), Early Herpetological Studies and Surveys in the Eastern United States. Arno Press, New York, NY].

Jensen, J.B., C.D. Camp, W. Gibbons, and M.J. Elliott. 2008. **Amphibians and Reptiles of Georgia**. University of Georgia Press, Athens, GA. xvii + 574 pp.

Laerm, J., B.J. Freeman, L.J. Vitt, J.M. Meyers, and L. Logan. 1980. Vertebrates of the Okefenokee Swamp. Brimleyana 4:47–73. [amphibians and reptiles on pp. 52–57].

Martof, B.S. 1956. **Amphibians and Reptiles of Georgia**. University of Georgia Press,

Athens, GA. viii + 94 pp.

Neill, W.T. 1948. The Lizards of Georgia. Herpetologica 4:153–158.

Neill, W.T. 1949. A Checklist of the Amphibians and Reptiles of Georgia. Ross Allen Reptile Institute, Silver Spring, FL. 4 pp.

Ruckdeschel, C., C.R. Shoop, and G.R. Zug. 2000. Sea Turtles of the Georgia Coast. Cumberland Island Museum, Cumberland Island, GA. x + 100 pp.

Wharton, C. 1953. Georgia Snake Tribe. Georgia State Museum of Science and Industry, Atlanta, GA. 11 pp. [revised edition 1963 – 22 pp.].

Williamson, G.K. and R.A. Moulis. 1979. Distribution of Georgia Amphibians and Reptiles in the Savannah Science Museum. Savannah Science Museum Special Publication No 1. 376 pp. (unpaginated).

Williamson, G.K. and R.A. Moulis. 1994. Herpetological Specimens in the Savannah Science Museum Collection. Volume 1. Amphibians, Volume 2 Reptiles. Savannah Science Museum Special Publication No 2. 726 pp.

Williamson, G.K. and R.A. Moulis. 1994. Distribution of the Amphibians and Reptiles in Georgia. Savannah Science Museum Special Publication No 3., 2 volumes, 912 pp.

Wright, A.H. 1932. Frogs of Okefinokee Swamp Georgia (North American Salienta No. 2). Macmillan, New York, NY. xv + 497 pp. [2002 facsimile reprint by Cornell University Press, xxiv + 509, with new foreword and afterword by J. W. Gibbons].

Wright, A.H. and W.D. Funkhouser 1915. A biological reconnaissance of the Okefinokee Swamp in Georgia – The reptiles. Proceedings of Academy of Natural Science of Philadelphia 67:108–192 + pls. 1–3.

Hawaii

Ching, P. 2001. **Sea Turtles of Hawai'i**. University of Hawai'i Press, Honolulu, HI. 55 pp.

Hunsaker, D. and P. Breese. 1967. Herpetofuana of the Hawaiian Islands. Pacific Science 21:423–428.

Jones, R.E. 1979. Hawaiian lizards: Their past, present, and future. Bulletin of Maryland Herpetological Society 15:37–45.

McGregor, R.C. 1904. Notes on Hawaiian reptiles from the Island of Maui. Proceedings of the United States National Museum 28:115–118.

McKeown, S. 1978. **Hawaiian Reptiles and Amphibians**. Oriental Publishing Co., Honolulu, HI. 80 pp.

McKeown, S. 1996. Field Guide to the Reptiles and Amphibians in the Hawaiian Islands. Diamond Head Publishers, Los Osos, CA. iv + 172 pp.

Oliver, J.A. and C.E. Shaw. 1953. The amphibians and reptiles of the Hawaiian Islands.

Zoologica 38:65–95.

Perkins, R.C.L. 1903. Vertebrata. pp. 365–466 *in* Fauna Hawaiensis, Being the Land–fauna of the Hawaiian Islands (or the Zoology of the Sandwich (Hawaiian) Isles). Vol 1(4). Cambridge University Press, Cambridge, England.

Snyder, J.O. 1917. Notes on Hawaiian lizards. Proceedings of the United States National Museum 54:19–25.

Stejneger, L. 1899. The land reptiles of the Hawaiian Islands. Proceedings of the United States National Museum 21:783–813.

Tinker, S.W. 1938. Animals of Hawaii, a Natural History of the Amphibians, Reptiles and Mammals Living in the Hawaiian Islands. The Nippu Jiji Co., Ltd., Honolulu, HI.188 pp. [amphibians and reptiles on pp. 13–71; revised 1941, Tongg Publishing Co., Honolulu, HI. 190 pp., amphibians and reptiles on pp. 15–73].

Tinker, S.W. 1980. A List of the Amphibians, Reptiles and Mammals of the Hawaiian Islands (exclusive of the Whales). S. W. Tinker, Honolulu, HI. 8 pp.

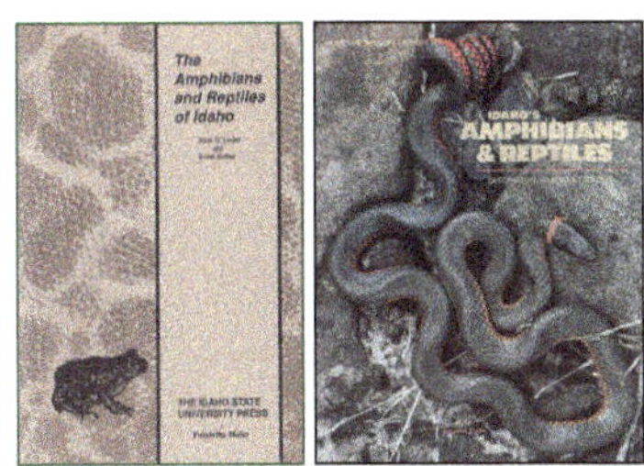

Idaho

Diller, L.V., and D.R. Johnson. 1982. Ecology of Reptiles in the Snake River Birds of Prey Area. Final Report. U.S. Department of Interior, Bureau of Land Management, Snake River Birds of Prey Research Project, Boise Idaho. 107 pp.

Fichter, E. and A.D. Linder. 1964. The Amphibians of Idaho. Idaho State University, Pocatello, Special Publication. 34 pp.

Groves, C. 1989. **Idaho's Amphibians and Reptiles: Description, Habitat and Ecology**. Nongame Wildlife Leaflet No. 7, Idaho Fish and Game, Boise, ID. 12 pp. [revised 1994].

Groves, C.R. and J.S. Marks. 1985. Annotated checklist of Idaho vertebrates. Tebiwa 22:10–27.

Koch, E.D. and C.R. Peterson. 1995. Amphibians and Reptiles of Yellowstone and Grand Teton National Parks. University of Utah Press, Salt Lake City, UT. xviii + 188 pp.

Lee, J.R. and C.R. Peterson. 2010. Herpetological Inventory of Craters of the Moon National Monument 1999–2001. National Park Service – Natural Resources Report, Fort Collins, CO. 133 pp.

Linder, A.D. and E. Fichter. 1970. The Reptiles of Idaho. Idaho State University Press Pocatello, ID. 45 pp.

Linder, A.D. and E. Fichter. 1977. **The Amphibians and Reptiles of Idaho**. Idaho State University Press, Pocatello, ID. 78 pp. [reprinted 1991].

Nussbaum, R.A., E.D. Brodie, Jr., and R.M. Storm. 1983. Amphibians and Reptiles of the Pacific Northwest. University Press of Idaho, Moscow, ID. 332 pp.

Peterson, C.R. 1994. Idaho's Amphibians and Reptiles: Description, Habitat and Ecology.

Nongame Wildlife Leaflet No. 7, Idaho Fish and Game, Boise, ID. 12 pp.

Shive, J.P. and C.R. Peterson. 2002. Herpetological Survey of Southcentral Idaho. Technical Bulletin No. 02–3. Idaho Bureau of Land Management, Boise, ID. 99 pp.

Slater, J.R. 1941. The distribution of amphibians and reptiles in Idaho. Occasional Papers of the Department of Biology, College of Puget Sound 14:78–109.

Tanner, W.W. 1941. The reptiles and amphibians of Idaho No. 1. Great Basin Naturalist 2:87–97.

Turner, F.B. 1955, Reptiles and Amphibians of Yellowstone Park. Yellowstone National Park, Yellowstone Interpretive Series No. 5. vi + 40 pp.

Van Denburgh, J. and J.R. Slevin. 1921. A list of the amphibians and reptiles of Idaho, with notes on the species in the collection of the Academy. Proceedings of the California Academy of Sciences, 4th Series, 11:39–47.

Illinois

Cagle, F.R. 1941. A Key to the Reptiles and Amphibians of Illinois. Contribution No. 5 of the Museum of Natural Science. Southern Illinois Normal University, Carbondale. iv + 32 pp., 3 pls.

Cahn, A.R. 1937. The Turtles of Illinois. University of Illinois Bulletin. 31. 218 pp., 31 pls.

Davis, N.S., Jr. and F.L. Rice. 1883. List of Batrachia and Reptilia of Illinois. Bulletin of Chicago Academy of Sciences. 1:25–32.

Edgren, R.A. and Stille, W. 1948. Checklist of Chicago area amphibians and reptiles. Natural History Misc. 26:1–7.

Garman, H. 1892. A synopsis of the reptiles and amphibians of Illinois. Illinois Laboratory of Natural History Bulletin 3:215–388, pls. 9–15.

Goodnight, C.J. 1937. A key to the adult salamanders of Illinois. Illinois State Academy of Science Transactions 30:300–302.

Kennicott, R. 1854. Catalogue of animals observed in Cook County, Illinois. Transactions of the Illinois State Agricultural Society 1853–54:577–612 [herpetological sections pp. 591–593].

Lueth, F.X. 1941. Manual of Illinois Snakes. Illinois Department of Conservation, Springfield. 48 pp. [revised 1949 (35 pp.) and 1956 (35 pp.); author's name spelled Leuth in revised editions].

Mauger, D. and T. G. Anton. 2015. Current distribution and status of amphibians and reptiles in Will County, Illinois. Illinois Natural History Survey Bulletin 40(1). iv + 31 pp.

Merczak, N. 1966. Snakes of Illinois. Bulletin of Chicago Herpetological Society 1(3):13–15.

Morris, M.A., R.S. Funk, and P.W. Smith. 1983. An annotated bibliography of the Illinois herpetological literature 1960–1980, and an updated checklist of species of the State. Illinois Natural History Survey Bulletin 33:123–138.

Necker, W.L. 1934. A synonymic catalogue of the reptiles and amphibians of Illinois. Illinois State Academy of Science Transactions 26:129.

Necker, W.L. 1938. Checklist of Reptiles and Amphibians of the Chicago Region. Chicago Academy of Sciences Leaflet No. 1. 4 pp.

Necker, W.L. 1939. Poisonous snakes of Illinois. The Chicago Naturalist 2:35–47

Necker, W.L. 1939. Revised Checklist of Reptiles and Amphibians of the Chicago Region. Chicago Academy of Sciences Leaflet No. 11. 4 pp.

Odd, D., Albert, S., Albert, K., Cravens, M., Harding, J. and Dennis, D. 2005. The Official Field Guide of Snakes of Chicagoland. Snakes of Chicagoland, Evanston, IL. 36 pp.

Parmalee, P.W. 1954. Amphibians of Illinois. Story of Illinois No. 10. Illinois State Museum, Springfield, IL. 38 pp.

Parmalee, P.W. 1955. Reptiles of Illinois. Illinois State Museum Popular Science Series 5. 88 pp.

Phillips, C.A., R.A Branson, and E.O. Moll. 1999. Field Guide to Amphibians and Reptiles of Illinois. Illinois Natural History Survey Manual 8. vx + 285 pp.

Phillips, C.A., J.A. Crawford, and A.R. Kuhns. 2022. Field Guide to Amphibians and Reptiles of Illinois, Second Edition. University of Illinois Press. Urbana, IL. 277 pp.

Pope, C.H. 1944. **Amphibians and Reptiles of the Chicago Area**. Chicago Natural History Museum. Chicago, IL. 275 pp. [reprinted 1947 and 1964].

Ross, H.H. 1964. Poisonous Snakes. Illinois Natural History Survey Reports 22. 4 pp.

Rossman, D.A. 1960. A herpetofaunal survey of the Pine Hills area of southern Illinois. Quarterly Journal of Florida Academy of Science 22:207–225.

Schmidt, K.P. 1929. The Frogs and Toads of the Chicago Area. Field Museum of Natural History Zoology Leaflet 11. 15 pp. + 5 pls.

Schmidt, K.P. 1930. The Salamanders of the Chicago Area. Field Museum of Natural History Zoology Leaflet 12. 16 pp. + 4 pls.

Schmidt, K.P. 1938. Turtles of the Chicago Area. Field Museum of Natural History Zoology Leaflet 14. 24 pp. + 2 pls.

Schmidt, K.P. and W.L. Necker. 1935. Amphibians and reptiles of the Chicago region. Bulletin of Chicago Academy of Science 5:57–77.

Smith, A.G. 1951. Key to the Amphibians and Reptiles of the Chicago Area. Contributions from the Department of Biological Sciences, Loyola University, Chicago, No.7. Leaflet No. 2. 4 pp.

Smith, P.W. 1947. The Reptiles and amphibians of eastern central Illinois. Bulletin of Chicago Academy of Sciences. 8:21–40.

Smith, P.W. 1953. Some facts about Illinois snakes and their control. State of Illinois, Natural History Survey Division, Biological Notes No. 32. 8 pp.

Smith, P.W. 1961. **The Amphibians and Reptiles of Illinois**. Illinois Natural History Survey Bulletin 28. 298 pp. [reprinted 1986].

Smith, P.W. and S.A. Minton, Jr. 1957. A distributional survey of the herpetofauna of Indiana and Illinois. American Midland Naturalist 58:341–351.

Vossler, J.J. 2021. Snake Road – Field Guide to the Snakes of Larue–Pine Hills. Southern Illinois University Press, Carbondale, IL. xiii + 152 pp.

Indiana

Blatchley, W.S. 1900. Notes on the batrachians and reptiles of Vigo County, Indiana. Annual Report of Indiana Department of Geology and Natural Resources 24:537–552.

Guljas, E. 1981. Understanding Indiana's Snakes. Indiana Department of Natural Resources Management Series No. 18. 11 pp.

Hay, O.P. 1887. A preliminary catalogue of the Amphibia and Reptilia of the State of Indiana. The Journal of the Cincinnati Society of Natural History 10(2):59–69.

Hay, O.P. 1887. The amphibians and reptiles of Indiana. Annual Report of the Indiana State Board of Agriculture 1887:201–223.

Hay, O.P. 1892. The batrachians and reptiles of the State of Indiana. Annual Report of the Indiana Geological Survey 17:410–602, 3 pls.

Hughes, E. 1886. Preliminary list of reptiles and batrachians of Franklin County. Bulletin of the Brookville Society of Natural History 2:40–45.

Karns, D.R. 1988. The herpetofauna of Jefferson County: analysis of an amphibian and reptile community in southeastern Indiana. Proceedings of the Indiana Academy of Sciences 95:535–552.

MacGowan, B. and B. Kingsbury. 2001. Snakes of Indiana. Indiana Nongame Program. West Lafayette, IN. 51 pp.

MacGowan, B. and R. Williams. 2012. Snakes and Lizards of Indiana. Purdue Extension West Lafayette, IN. 96 pp.

MacGowan, B. and R. Williams. 2013. Appreciating Reptiles and Amphibians in Nature Purdue Extension, West Lafayette, IN. 20 pp.

MacGowan, B., B. Kingsbury and R. Williams. 2005. Turtles of Indiana. Purdue Extension, West Lafayette, IN. 63 pp.

Minton, Jr., S.A. 1944. Introduction of the study of the reptiles of Indiana. American Midland Naturalist. 32:438–477.

Minton, Jr., S.A. 1966. Amphibians and Reptiles. pp. 426–451 in Lindsey, A.A. (ed.), Natural Features of Indiana. Indiana Academy of Science, Indianapolis, IN.

Minton, Jr., S.A. 1972. **Amphibians and Reptiles of Indiana**. Monograph Indiana Academy of Science No. 3. v + 346 pp.

Minton, Jr., S.A. 2001. **Amphibians and Reptiles of Indiana**. Indiana Academy of Sciences, Indianapolis, IN. xiv + 404 pp.

Mittleman, M.B. 1947. Miscellaneous notes on Indiana amphibians and reptiles. American Midland Naturalist 38:466–484.

Myers, G.S. 1926. A synopsis of the identification of the amphibians and reptiles of Indiana. Proceedings of the Indiana Academy of Science 34:277–294.

Pope, C.H. 1947. Amphibians and Reptiles of the Chicago Area. Chicago Natural History Museum, Chicago, IL. 275 pp.

Schmidt, K.P. 1929. The Frogs and Toads of the Chicago Area. Field Museum of Natural History Zoology Leaflet 11. 15 pp. + 5 pl.

Schmidt, K.P. 1930. The Salamanders of the Chicago Area. Field Museum of Natural

History Zoology Leaflet 12. 16 pp. + 4 pl.

Schmidt, K.P. 1938. Turtles of the Chicago Area. Field Museum of Natural History Zoology Leaflet 14. 24 pp. + 2 pl.

Schmidt, K.P. and W.L. Necker. 1935. Amphibians and reptiles of the Chicago region. Bulletin of Chicago Academy of Science 5:57–77.

Smith, P.W. and S.A. Minton, Jr. 1957. A distributional survey of the herpetofauna of Indiana and Illinois. American Midland Naturalist 58:341–351.

Williams, R.N., B.J. MacGowan, B. Kingsbury, and Z. Walker. 2006. Salamanders of Indiana. Purdue Extension, West Lafayette, IN. 91 pp.

Williams, R.N., B.J. MacGowan, and Z. Walker, J. Hoverman, and N. Burgmeier. 2017. Frogs and Toads of Indiana. Purdue Extension, West Lafayette, IN. 49 pp.

Iowa

Bailey, R.M. and M.K. Bailey. 1941. The distribution of Iowa toads. Iowa State College Journal of Science 15:169–177.

Bailey, R.M. 1944. Iowa's frogs and toads. Iowa Conservationist 3(3):17–20 & 3(4):25, 27–30.

Christiansen, J.L. and R.M. Bailey. 1986. **The Snakes of Iowa**. Iowa Conservation Commission, Des Moines, IA. 15 pp. [revised 1990].

Christiansen, J.L. and R.M. Bailey. 1988. The Lizards and Turtles of Iowa. Nongame Technical Series No. 2, Iowa Conservation Commission, Des Moines, IA. 19 pp.

Christiansen, J.L. and R.M. Bailey. 1991. The Salamanders and Frogs of Iowa. Nongame Technical Series No. 3, Iowa Conservation Commission, Des Moines, IA. 24 pp.

Christiansen, J.L. and R.R. Burken. 1978. The Endangered and Uncommon Reptiles and Amphibians of Iowa. Iowa State Science Teachers Journal (special issue). 26 pp.

Geske, J. 1998. Iowa Reptiles and Amphibians. Iowa Association of Naturalists. Guthrie Center, IA. 24 pp.

Guthrie, J.E. 1926. The snakes of Iowa. Iowa State Agricultural Experiment Station Bulletin 239:147–192.

LeClere, J. 1998. Checklist of the herpetofauna of Iowa. Minnesota Herpetological Society Occasional Paper Number 5. 23 pp.

LeClere, J. 2013. **A Field Guide to the Amphibians and Reptiles of Iowa**. ECO Publishing, Rodeo, NM. viii + 350 pp.

Parmelee, J.R., M.G. Knutson, and J.E. Lyon. 2002. A Field Guide to Amphibian Larvae and Eggs of Minnesota, Wisconsin, and Iowa. U.S. Geological Survey Information and Technology Report 2002–0004. 38 pp.

VanDeWalle, T. 2022. The Natural History of the Snakes and Lizards of Iowa. University of Iowa Press, Iowa City, IA. 379 pp.

Kansas

Ashton, R.E., Collins, J.T. and Ferguson, G.W. 1974. Reptiles and Amphibians (Leaders Guide). Cooperative Extension Service, Kansas State University, Manhattan, KS. 4–H 371. 19 pp. [note: Leaders Guide has keys and checklist, lacking in the corresponding Members Manual].

Branson, E.B. 1904. Snakes of Kansas. Kansas University Science Bulletin 2:353–430.

Burt, C.E. 1928. The lizards of Kansas. Transactions of the Academy of Science of St. Louis 26(1):1–85.

Busby, W.H., J.R. Parmelee, C.M. Dwyer, E.D. Hooper, and K.J. Irwin. 1994. A survey of the herpetofauna of the Fort Riley Military Reservation, Kansas. State Biological Survey of Kansas, Report 58. iv + 79 pp.

Busby, W.H., J.T. Collins, and J.R. Parmelee. 1996. Reptiles and Amphibians of Fort Riley and Vicinity. Kansas Biological Survey, Topeka, KS. vi + 72 pp.

Busby, W.H., J.T. Collins, G. Suleiman. 2005. The Snakes, Lizards, Turtles, and Amphibians of Fort Riley and Vicinity. Kansas Biological Survey, Topeka, KS. iiv + 76 pp.

Caldwell, J.P. and J.T. Collins. 1981. Turtles in Kansas. AMS Publishing, Lawrence, KS. x + 67 pp.

Clarke, R.F. 1956. Turtles in Kansas. Kansas School Naturalist 2(4):1–15.

Clarke, R.F. 1959. Poisonous Snakes of Kansas. Kansas School Naturalist 5(3):1–16.

Clarke, R.F. 1965. Lizards in Kansas. Kansas School Naturalist 11(4):1–16.

Clarke, R.F. 1970. Salamanders in Kansas and Vicinity. Kansas School Naturalist 16(4):1–16.

Clarke, R.F. 1980. Snakes in Kansas. Kansas School Naturalist 26(3):1–15.

Clarke, R.F. 1984. Frogs and toads in Kansas. Kansas School Naturalist 30(3):1–15.

Collins, J.T. 1974. Amphibians and Reptiles in Kansas. University of Kansas, Museum of Natural History Public Education Series No. 1. 283 pp.

Collins, J.T. 1976. Kansas snakes. An educational experience. Kansas Fish and Game 33(4):16–19.

Collins, J.T. 1977. Kansas frogs and toads. Kansas Fish and Game 34(3):12–16.

Collins, J.T. 1980. Kansas turtles. Kansas Fish and Game 37(3):11–14.

Collins, J.T. 1982. Amphibians and Reptiles in Kansas. University of Kansas, Museum of Natural History Public Education Series No. 8, second edition. xiii + 356 pp.

Collins, J.T. 1984. The lizards of Oz. Kansas Wildlife 41(2):12–16.

Collins, J.T. 1993. Amphibians and Reptiles in Kansas. University of Kansas, Museum of Natural History Public Education Series No.13, third edition. xx + 397 pp.

Collins, J.T. 1996. A Guide to the Great Snakes of Kansas. Kansas Nature Guides, Lawrence, KS. 33 pp. [reprinted 1998, revised 2001, 29 pp.].

Collins, J.T. and J.P. Caldwell. 1977. A bibliography of the amphibians and reptiles of Kansas (1854–1976). Report of State Biological Survey [Kansas] 12:1–56.

Collins, J.T. and S.L. Collins. 1991. Reptiles and Amphibians of the Cimarron National

Grasslands, Morton County, Kansas. US Forest Service, Lawrence, KS. viii + 60 pp. [revised 2011, viii + 67 pp.].

Collins, J.T. and S.L. Collins. 1993. Reptiles and Amphibians of Cheyenne Bottoms. Hearth Publ., Hillsboro, KS. xii + 92 pp.

Collins, J.T. and S.L. Collins. 2006. A Pocket Guide to Kansas Snakes. Friends of Great Plains Nature Center, Wichita, KS. 69 pp. [2nd edition 2009, 3rd edition 2010, 6th edition 2017, others not seen].

Collins, J.T. and S.L. Collins. 2006. Amphibians, Turtles, and Reptiles of Cheyenne Bottoms, second edition. US Fish and Wildlife, Fort Hays, KS. viii + 76 pp.

Collins, J.T, S.L. Collins and T.W. Taggart. 2010. **Amphibians, Reptiles, and Turtles in Kansas**. Eagle Mountain Publishing, Eagle Mountain, UT. xvi + 312 pp.

Collins, J.T., P. Gray, H. Guarisco, K.J. Irwin, and L. Miller. 1981. The Kansas Herpetological Society Presents Endangered and Threatened Amphibians and Reptiles in Kansas. Kansas Herpetological Society, Lawrence, KS. 4 pp.

Cragin, F.W. 1881. A preliminary catalogue of Kansas reptiles and amphibians. Transactions of the Kansas Academy of Science 7:112–120.

Fitch, H.S. 1999. A Kansas Snake Community: Composition and Changes Over 50 Years. Krieger Publishing Company, Malabar, Florida. 165 pp.

Heinrich, M.L. and D.W. Kaufman. 1985. Herpetofauna of the Konza Prairie Research Natural Area, Kansas. Prairie Naturalist 17:101–112.

Karns, D., R.E. Ashton, Jr., and T. Swearingen. 1974. Illustrated Guide to Amphibians and Reptiles in Kansas. University of Kansas, Museum of Natural History Public Education Series No. 2. 28 pp.

Knight, J.L. and J.T. Collins. 1977. The Amphibians and Reptiles of Cheyenne County, Kansas. Biological Survey of Kansas Report 15, Lawrence, KS. 18 pp.

Platt, D.R. 1973. Rare, endangered, and extirpated species in Kansas II: Amphibians and reptiles. Transactions of the Kansas Academy of Science 76:185–192.

Rundquist, E.M. and Collins, J.T. 1977. The amphibians of Cherokee County, Kansas. Reports of the State Biological Survey of Kansas 14. [1] + 12 pp.

Smith, H.M. 1934. The amphibians of Kansas, American Midland Naturalist 15:377–528.

Smith, H.M. 1950. **Handbook of Amphibians and Reptiles of Kansas**. University of Kansas Museum of Natural History Miscellaneous Publication No. 2. 336 pp.

Smith, H.M. 1956. Handbook of Amphibians and Reptiles of Kansas, second edition. University of Kansas Museum of Natural History Miscellaneous Publication No. 9 356 pp.

Taggart, T.W. and J.D. Riedle. 2017. A Pocket Guide to Kansas Amphibians, Turtles, and Lizards. Great Plains Nature Center, Wichita, KS. 69 pp.

Taylor, E.H. 1929. A revised checklist of the snakes of Kansas. University of Kansas Science Bulletin 19:53–62.

Taylor, E.H. 1993. The Lizards of Kansas. Kansas Herpetological Society Special Publication No.2. 72 pp. [published version of Taylor's 1916 MS Thesis].

Kentucky

Barbour, R.W. 1956. The Salientia of Kentucky: Identification and distribution. Transactions of the Kentucky Academy of Science 17:81–86.

Barbour, R.W. no date [1952 or later]. A Checklist and Key to the Amphibians and Reptiles of Kentucky. University of Kentucky, Lexington, KY. [ii] + 37 pp. [revised 1957, [ii] + 41 pp. + 6 pp. data sheets; additional editions or reprints distributed as class material].

Barbour, R.W. 1971. **Amphibians and Reptiles of Kentucky**. University Press of Kentucky, Lexington, KY. x + 334 pp.

Barnes, T. 1999. Snakes: Information for Kentucky Homeowners. University of Kentucky Extension, Lexington, KY. 6 pp.

Collins, J.T. 1964. A preliminary review of the snakes of Kentucky. Journal of the Ohio Herpetological Society 4: 69–77.

Collins, J.T. and R.E. Kurtz. 1977. A Bibliography of the Amphibians and Reptiles of Kentucky (1820–1976). Kentucky Academy of Sciences, Lawrence, KS. 21 pp.

Dourson, D. and J. Dourson. 2021. Reptiles and Amphibians of Red River Gorge, Second Edition. Goatslug Publications, Stanton, KY. 204 pp. [2nd edition 2022 (210 pp.)].

Dury, R. and R.S. Williams 1933. Notes on some Kentucky amphibians and reptiles. Bulletin of Baker–Hunt Museum 1:1–22.

Ernst, C.H. and E.M. Ernst. 1969. Turtles of Kentucky. International Turtle and Tortoise Society Journal 3(5):13–15.

Funkhouser, W.D. 1925. Wildlife in Kentucky. State of Kentucky, Frankfort, KY. 385 pp. [reptiles on pages 61–141].

Funkhouser, W.D. 1945. Kentucky Snakes. Department of Extension, University of Kentucky, Lexington, KY. 31 pp.

Garman, H. 1894. A preliminary list of the Vertebrate animals of Kentucky. Bulletin of the Essex Institution 26:1–63. [herpetology on pp. 34–39, 61–63.].

Green River Lake Staff. 1978. Poisonous Snakes of Central Kentucky. US Army Corps of Engineers, Louisville, KY. 12 pp.

Mead, L. 1991. Kentucky snakes: Their systematics, variation, and distribution. Journal of the Tennessee Academy of Sciences 66:175–182.

Meade, L.E. 2005. Kentucky Snakes: Their Identification, Variation and Distribution. Kentucky Nature Preserves Commission, Frankfort, KY. xiv + 323 pp. + 17 pl.

Moore, B. and T. Slone. 2002. **Kentucky Snakes**. Kentucky Department of Fish and Wildlife. Frankfort, KY. 32 pp. [reprinted 2009, 2010, 2015 and possibly others].

Snyder, D.H. 1972. Amphibians and Reptiles of Land Between the Lakes. Tennessee Valley Authority, Golden Pond, KY. 90 pp.

Snyder, D.H., A.F. Scott, E. Zimmerer, and D. Frymire. 2016. Amphibians and Reptiles of Land Between the Lakes. University Press of Kentucky, Lexington, KY. 107 pp.

Welter, W.A. and K. Carr. 1939. Amphibians and reptiles of northeastern Kentucky. Copeia 1939:128–130.

Louisiana

Behre, E.H. 1950. Annotated List of the Fauna of the Grand Isle Region 1928–1946. Occasional Papers of the Marine Laboratory of Louisiana State University 6. 66 pp.

Beyer, G.E. 1900. Louisiana herpetology, with a check-list of the batrachians and reptiles of the State. Proceedings of the Louisiana Society of Naturalists. 1897–1899:25–46.

Bick, G.H. 1954. A bibliography of the zoology of Louisiana. Proceedings of the Louisiana Academy of Sciences 17: 15–48.

Boundy, J. 1997. Snakes of Louisiana. Louisiana Department of Wildlife and Fisheries, Baton Rouge, LA. 32 pp.

Boundy, J. 2006. Snakes of Louisiana, Second Edition. Louisiana Department of Wildlife and Fisheries, Baton Rouge, LA. 40 pp.

Boundy, J. and J.L. Carr. 2017. **Amphibians and Reptiles of Louisiana**. Louisiana State University Press, Baton Rouge, LA. 386 pp.

Cagle, F.R. 1952. A Key to the Amphibians and Reptiles of Louisiana. Tulane University, New Orleans, LA. v + 42 pp.

Clark, R.F. 1949. Snakes of the hill parishes of Louisiana. Journal of Tennessee Academy of Sciences 24(4):244–261.

Dundee, H.A. and D.A. Rossman. 1989. **The Amphibians and Reptiles of Louisiana**. Louisiana State University Press, Baton Rouge, LA. xi + 300 pp. [softcover reprint 1996].

Gowanloch, J. N. 1934. Poisonous snakes of Louisiana. Louisiana Conservation Review 4(3): 1–16.

Gowanloch, J.N. 1943. Poisonous snakes of Louisiana. pp. 8–62 *in* Gowanloch, J.N. and C.A. Brown. Poisonous Snakes, Plants, and Black Widow Spider of Louisiana. Louisiana Department of Conservation, New Orleans, LA. 133 pp.

Hardy, L.M. 1979. Checklist of the Amphibians and Reptiles of Caddo and Bossier Parishes, Louisiana. Bulletin of Museum of Life Science, Louisiana State University 2. 11 pp.

Hardy, L.M. 1995. Checklist of the Amphibians and Reptiles of Caddo Lake Watershed of Texas and Louisiana. Bulletin of Museum of Life Science, Louisiana State University 10. 31 pp.

Keiser, Jr., E.D. and L.D. Wilson. 1969. Checklist and Key to the Herpetofauna of Louisiana. Lafayette Natural History Museum Technical Bulletin No. 1. [iv] + 51 pp. [2nd edition1979, [iv] + 49 pp.].

Keiser, Jr., E.D. 1971. The poisonous snakes of Louisiana and the emergency treatment of their bites. Louisiana Conservationist 23(7/8): 1–16.

Liner, E.A. 1954. The herpetofauna of Lafayette, Terrebonne, and Vermillion Parishes, Louisiana. Proceedings of the Louisiana Academy of Science 17:65–85.

Strecker, J.K. and L.S. Frierson, Jr. 1926. The herpetology of Caddo and DeSoto parishes, Louisiana. Contributions from Baylor University Museum 5. 8 pp.

Viosca, Jr., P. 1931. Amphibians and Reptiles of Louisiana. Southern Biological Supply Co., Incorporated. Price List No. 20. Herpetology. 12 pp.

Viosca, Jr., P. 1950. Amphibians and Reptiles of Louisiana. Popular Science Bulletin No.1. Louisiana State Univ. Extension. 12 pp. [reprinted 1965].

Viosca, Jr., P. 1960. Salamanders in Louisiana. Louisiana Conservationist 12(1):8–9, 15–16. [reprinted as Wildlife Education Bulletin No. 36. Louisiana Wild Life and Fisheries Commission, New Orleans. 11 pp.].

Viosca, Jr., P. 1960. Frogs and toads of Louisiana. Louisiana Conservationist 12(2–3): 8–9, 20. [reprinted as Louisiana Wild Life and Fisheries Commission, New Orleans. 7 pp.].

Viosca, Jr., P. 1960. Poisonous snakes of Louisiana. Louisiana Conservationist 13(11–12):2–3, 24. [actual volume number is 12; reprinted as Wildlife Education Bulletin No.9. Louisiana Wild Life and Fisheries Commission, New Orleans. 8 pp., multiple Reprints through at least 1968].

Viosca, Jr., P. 1961. Harmless snakes of Louisiana. Louisiana Conservationist 13(2):17–19, 23. [reprinted as Wildlife Education Bulletin No. 10. Louisiana Wild Life and Fisheries Commission, New Orleans. 8 pp. [revised 1962 (8 pp.), 1966 (8 pp.), 1975 (10 pp.)].

Viosca, Jr., P. 1961. Turtles, tame and truculent. Louisiana Conservationist 13(7–8):5–8. [reprinted as 12 pp. pamphlet by Louisiana Wild Life and Fisheries Commission, no date].

Viosca, Jr., P. 1980. Poisonous Snakes and Harmless Snakes of Louisiana. Baton Rouge, Louisiana Department of Wildlife and Fisheries. Wildlife Education Bulletins No. 9 & No. 10. [2] + 17 pp.

Walker, J. M. 1963. Amphibians and reptiles of Jackson Parish, Louisiana. Proceedings of the Louisiana Academy of Sciences 26:91–101.

Maine

Andrews, P. 1966. Turtles. Maine Fish and Game 7(3):20.

Fogg, B.F. 1862. List of reptiles and amphibians found in the State of Maine. Proceedings of the Portland Society of Natural History 1:86.

Hunter, M. L. 1984. A Preliminary Guide to Finding and Identifying the Amphibians and Reptiles of Maine. The Nature Conservancy, Topsham, ME. 45 pp.

Hunter, M. L., J. Albright, and J. Arbuckle (eds). 1992. **The Amphibians and Reptiles of Maine**. Maine Agricultural Experiment Station Bulletin 838. 188 pp.

Hunter, Jr., M. L., A.J. Calhoun, and M. McCollough. 1999. **Maine Amphibians and Reptiles**. University of Maine Press, Orono, ME. 252 pp. + CD.

Lowe, R. 1928. Snakes of Maine. Maine Naturalist 8:26–29.

Mairs, D.F. 1965. Salamanders. Maine Fish and Game 7(1):20–22.

Mairs, D.F. 1968. Maine's frogs and toads – The swampland singers. Maine Fish and Game, 10(1):16–18

McCollough, M. 1997. Status and conservation of turtles in Maine. pp. 7–11 *in* T.F. Tyning, (ed.), Status and Conservation of Turtles of the Northeastern United States. Serpent's Tale Natural History Books, Lanesboro, MN. ix+53 pp.

Mincher, W.L. 1971. Maine snakes are harmless. Maine Fish and Game 13(3):20–23.

Verrill, A.E. 1863. Catalogue of the reptiles and batrachians found in the vicinity of Norway, Oxford Co., Maine. Proceedings of the Boston Society of Natural History 9:195–199.

Maryland and Washington D.C.

Bierly, E.J. 1954. Turtles in Maryland and Virginia. Atlantic Naturalist 9:244–249.

Cohen, E. 1936. A key to Maryland turtles we should know. Natural History Society of Maryland Junior Bulletin 1(2):4–6.

Conant, R. 1945. An Annotated Checklist of the Amphibians and Reptiles of the Del-Mar-Va Peninsula. Society of Natural History of Delaware, Wilmington, DE. 8 pp.

Cooper, J.E. 1948. Maryland frogs and toads. Maryland Naturalist 18:37–40.

Cooper, J.E. 1960. Distributional survey of Maryland and District of Columbia. Bulletin of Philadelphia Herpetological Society 8(3):18–24.

Cooper, J.E. 1965. Distributional survey: Maryland and District of Columbia. (Revised by H.S. Harris, Jr.) Bulletin of Maryland Herpetological Society 1:3–14.

Cunningham, H.R, and N.H. Nazdrowicz. 2018. **The Maryland Amphibian and Reptile Atlas**. John Hopkins University Press, Baltimore, MD. 283 pp.

Hardy, Jr., J.D. 1972. Amphibians of the Chesapeake Bay region. Chesapeake Science 13 (supplement):123–128.

Hardy, Jr., J.D. 1972. Reptiles of the Chesapeake Bay region. Chesapeake Science 13 (supplement):128–134.

Harris, Jr., H.S. 1969. Distributional survey: Maryland and the District of Columbia. Bulletin of Maryland Herpetological Society 5:97–161.

Harris, Jr., H.S. 1975. Distributional survey (Amphibia / Reptilia). Maryland and District of Columbia. Bulletin of Maryland Herpetological Society 11:73–167.

Hay, W.P. 1902. A list of the batrachians and reptiles of the District of Columbia and vicinity. Proceedings of the Biological Society of Washington 15:121–145.

Kelly, H.A., A.W. Davis, and H.C. Robertson. 1936. Snakes of Maryland. Natural History Society of Maryland, Hagerstown, MD. 103 pp.

Klimkiewicz, M.K. 1972. Reptiles of Mason Neck. Atlantic Naturalist 27:20–25.

Mansueti, R. 1941. A descriptive catalogue of the amphibians and reptiles found in and around Baltimore City, Maryland, within a radius of twenty Miles. Proceedings of the Natural History Society of Maryland 7:1–53 + 2 pls.

Mansueti, R. 1942. Notes on the herpetology of Calvert County, Maryland. Bulletin of the Natural History Society of Maryland 12:33–43, 1 pl.

Mansueti, R. 1947. Maryland salamanders. Maryland Naturalist 17(4):81–84.

Maryland Herpetological Society. 1973. Endangered amphibians and reptiles of Maryland. Bulletin of the Maryland Herpetological Society 9: 42–100.

McCauley, Jr., R. H. 1945 The Reptiles of Maryland and District of Columbia. Natural History Society of Maryland, Hagerstown, MD. 194 pp.

McClellan, W.H., R. Mansueti, Francis Groves. 1943. The lizards of central and southern Maryland. Proceedings of the Natural History Society of Maryland 8:1–41 + 8 pls.

Norman, J.E. 1949. Maryland turtles. Maryland Naturalist 19:13–16.

Reed, C.F. 1956. Contributions to the Herpetology of Maryland and Delmarva, 5. Bibliography to the Herpetology of Maryland, Delmarva, and the District of Columbia. Reed Herpetorium (privately published) 9 pp.

Reed, C.F. 1956. Contributions to the Herpetology of Maryland and Delmarva, 6. An Annotated Check List of the Lizards of Maryland and Delmarva. Reed Herpetorium (privately published), Baltimore, MD. 6 pp.

Reed, C.F. 1956. Contributions to the Herpetology of Maryland and Delmarva, 7. An Annotated Check List of the Turtles of Maryland and Delmarva. Reed Herpetorium (privately published), Baltimore, MD. 11 pp.

Reed, C F. 1956. Contributions to the Herpetology of Maryland and Delmarva, 8. An Annotated Check List of the Snakes of Maryland and Delmarva. Reed Herpetorium (privately published), Baltimore, MD. 20 pp.

Reed, C.F. 1956. Contributions to the Herpetology of Maryland and Delmarva, 9. An Annotated Check List of the Frogs and Toads of Maryland and Delmarva. Reed Herpetorium (privately published), Baltimore, MD. 19 pp.

Reed, C.F. 1956. Contributions to the Herpetology of Maryland and Delmarva, 11. An Annotated Herpetofauna of the Delmarva Peninsula, including Many New and Additional Localities. Reed Herpetorium (privately published), Baltimore, MD. 11 pp. + 1 not numbered.

Reed, C.F. 1957. Contributions to the Herpetology of Maryland and Delmarva, 10. An Annotated Check List of the Salamanders of Maryland and Delmarva. Reed Herpetorium (privately published), Baltimore, MD. 6 pp.

Reed, C.F. 1957. Contributions to the herpetology of Maryland and Delmarva, 15: The herpetofauna of Somerset County, Md. Journal of the Washington Academy of Sciences 47:127–128.

Roberts, D.A. 1957. A Simplified Guide to Native Snakes of Maryland. Donald A. Roberts, Baltimore. [4] + 8 pp.

Schwartz, F.J. 1961. **Maryland Turtles**. Bulletin of Maryland Department of Education 50:1–44.

Schwartz, F.J. 1967. Maryland Turtles. University of Maryland, Natural Resources Insti-

tute, Educational Series 79. 38 pp.
White, Jr., J.F. and A.W. White. 2002. Amphibians and Reptiles of Delmarva. Tidewater Publishers, Centreville, MD. xvi + 248 pp., 32 pp. pls.
White, Jr., J. F., and A. W. White. 2007. Amphibians and Reptiles of Delmarva. Second Edition. Delaware Nature Society, Inc., Tidewater Publishers, Centreville, Maryland. xvi + 243 pp., 32 pp. pls. [accompanied by an addendum sheet dated 2008, reprinted 2009].

Massachusetts

Allen, J.A. 1868. **Catalogue of the reptiles and batrachians found in the vicinity of Springfield, Massachusetts with notes of all other species known to inhabit the state.** Proceedings of the Boston Society of Natural History 12:171–204, 248–250.
Allen, J.A. 1870. Notes on Massachusetts reptiles and batrachians. Proceedings of the Boston Society of Natural History Society 13:260–263.
Babbitt, L.H. and T.E. Graham. 1972. Snakes of Massachusetts. Massachusetts Wildlife 23(6):7–9,14–19.
Blazis, M.M. and Underwood, K.J. (eds.). 1991. A Field Guide to the Amphibians and Reptiles of Auburn and Southern Worcester County. The Auburn Middle School Field Guide Series, Volume 3. x + 85 + [2].
Cutson, J. 1994. A Guide to Amphibians and Reptiles of the Greater Amherst Area. Jaana Cutson, Amherst, MA. 11 pp.
Dunn, E.R. 1930. Reptiles and Amphibians of Northhampton and Vicinity. Boston Society of Natural History Bull 57:3-8.
Graham, T.E. 1970. Sportsman's guide to Massachusetts freshwater turtles. Massachusetts Wildlife 21(1):8–13.
Graham, T.E. 1978. The salamanders of Massachusetts. Massachusetts Wildlife 29(4):2–9, 15–16.
Graham, T.E. 1978. Massachusetts frogs and toads. Part I. Massachusetts Wildlife 29(5):12–14.
Graham, T.E. 1978. Massachusetts frogs and toads. Part II. Massachusetts Wildlife 29(6):12–19.
Greer, A.E., T.S. Doyle, and P Arnold. 1973. Annotated Checklist of the Amphibians and Reptiles of Concord, Carlisle, and Bedford, Massachusetts. Concord Field Station, Harvard. 21 pp.
Herscu, M. 2004. Snakes of the Pioneer Valley. AppleBlossom Press, Amherst, Massachusetts. 144 pp.
Jackson, S. and P. Mirick. n.d. [2001]. Massachusetts Snakes: A Guide. Massachusetts Division of Fisheries and Wildlife, Westborough, MA. 20 pp.

Kenney, L.P. and M.R. Burne. 2000. A Field Guide to the Animals of Vernal Pools. Massachusetts Division of Fish and Wildlife, Westborough MA. 73 pp.

Lazell, Jr., J.D. 1972. Reptiles and Amphibians in Massachusetts. Massachusetts Audubon Society, Lincoln, MA. 35 pp. [revised 1974].

Lazell, Jr., J.D. 1976. This Broken Archipelago. Quadrangle, New York., NY. xi + 260 pp. Massachusetts Division of Fish and Wildlife. 2009. Field Guide to the Reptiles of Massachusetts. Massachusetts Wildlife 59(2):1–45.

Massachusetts Division of Fish and Wildlife 2013. Field Guide to the Amphibians of Massachusetts. Massachusetts Wildlife 59(2):1–37. [volume number incorrectly printed in magazine; correct volume is 63].

Mirick, P.G., T. French, and J. Kubel. 2016. **Field Guide to the Amphibians and Reptiles of Massachusetts**. Massachusetts Division of Fisheries and Wildlife, Westborough, MA. 94 pp.

Smith, D.S. 1833. III Reptilia or Reptiles. P. 552 *in* Hitchcock, E. Report of the Geology, Mineralogy, Botany, and Zoology of Massachusetts. Press of J.S. and C. Adams. Amherst, MA. 702 pp.

Storer, D. H. 1839. A report on the reptiles of Massachusetts. pp. 203-253 in Reports on the Fishes, Reptiles and Birds of Massachusetts. Commission on the Zoological and Botanical Survey of the State (Massachusetts), Boston. xv + 425 pp. [reprinted 1978 *in* Adler, K. (ed.), Early Herpetological Studies and Surveys in the Eastern United States. Arno Press, New York, NY].

Storer, D. H. 1840. A report on the reptiles of Massachusetts. Boston Journal of Natural History 3:1–64 + pl. 4.

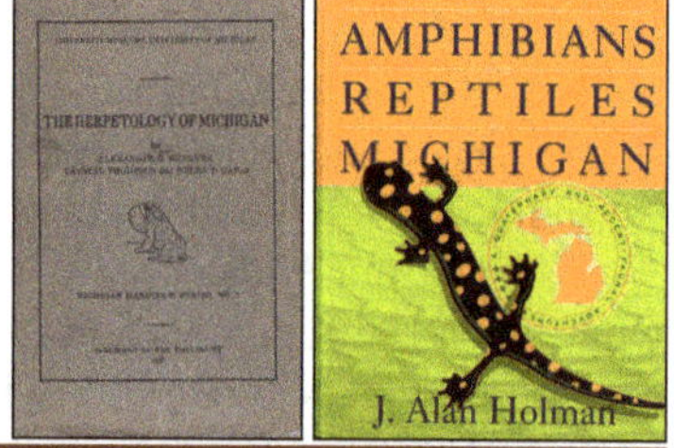

Michigan

Blanchard, N.F. 1928. Amphibians and reptiles of the Douglas Lake region in northern Michigan. Copeia 167:42–51.

Carpenter, C.C. 1957. Michigan's turtles and lizard. The Jack-Pine Warbler 35:82–85.

Carpenter, C.C. 1957. Michigan's salamanders. The Jack-Pine Warbler 35:6–8.

Carpenter, C.C. 1957. Michigan's snakes. The Jack-Pine Warbler 35:115–117.

Carpenter, C.C. 1957. Michigan's frogs and toads. The Jack-Pine Warbler 35:64–67.

Casper, G. S. 2005. An Amphibian and Reptile Inventory of Pictured Rocks National Lakeshore. U.S. Department of the Interior, National Park Service, Natural Resource Program Center, Natural Resource Technical Report NPS/GLKN/NRTR–2008/146. 48 pp.

Casper, G. S. 2008. An Amphibian and Reptile Inventory of Isle Royale National Park. U.S. Department of the Interior, National Park Service, Natural Resource Program Center, Natural Resource Technical Report NPS/ GLKN/2005/05. 41 pp.

Ellis, M.M. 1917. Amphibians and reptiles of Douglas Lake (Michigan) region. Michigan

Academy of Sciences Report 19:45–63.
Gibbs, M., F.N. Notestein, and H.L. Clark. 1905. A preliminary list of the amphibians and reptiles of Michigan. Annual Report of the Michigan Academy of Science 7:109–110.
Harding, J.H. 1989. A time for turtles. Michigan Natural Resources Magazine 58(3):36–43.
Harding, J.H. and J.A. Holman. 1990. Michigan Turtles and Lizards. Michigan State University Cooperative Extension Service, East Lansing, MI. 94 pp.
Harding, J.H. and J.A. Holman. 1992. Michigan Frogs, Toads, and Salamanders. Michigan State University Cooperative Extension Service, East Lansing, MI. 144 pp.
Hensley, M.M. 1975. An Illustrated Key to the Amphibians of Michigan. Woldumar Nature Center, Lansing, MI. 16 pp.
Holman, J.A. 2012. **Amphibians and Reptiles of Michigan**. Wayne State University Press, Detroit, MI. xix + 290 pp.
Holman, J.A. and J.H. Harding. 1977. Michigan's Turtles. Publication of the Museum, Michigan State University, Education Bulletin 3. 40 pp.
Holman, J.A., and J.H. Harding. 2006. Michigan Snakes. A Field Guide and Pocket Reference. Revised Edition. Michigan State University, East Lansing, MI. 74 pp.
Holman, J.A., J.H. Harding, M.M. Hensley, and G.R. Dudderar. 1989. Michigan Snakes. Michigan State University Cooperative Extension Service, East Lansing, MI. 72 pp.
Lagler, K.F. 1954. Michigan turtles. Michigan Conservation, May–June 1954:15–19.
Manville, R.H. 1954. The Snakes of Michigan. Michigan State College Extension Bulletin 315. 20 pp.
Mifsud, D. 2014. Michigan Amphibian and Reptile Best Management Practices. Herpetological Resources and Management, Chelsea, MI. 163 pp.
Miles, M. 1861. A Catalogue of the Mammals, Birds, Reptiles, and Molluscs of Michigan. First Biennial Report of the Progress of the Geological Survey of Michigan. Pp. 219–242. [amphibians and reptiles on pp. 22–25].
Notestein, F.N. 1905. The Ophidia of Michigan. Annual Report of Michigan Academy of Science. 7:111–125.
Ruthven, A.G. 1906. The cold–blooded vertebrates of the Porcupine Mountains and Isle Royale, Michigan. Report of the Geological Survey of Michigan for 1905: 107–112.
Ruthven, A.G. 1909. The cold–blooded vertebrates of Isle Royale. Biological Survey of Michigan 1909:329–333.
Ruthven, A.G., C. Thompson, and H. Thompson. 1912. The Herpetology of Michigan. Michigan Geological and Biological Survey 10(3):1–166 + pls. 1–20.
Ruthven, A.G., C. Thompson, and H.T. Gaige. 1928. **The Herpetology of Michigan**. University of Michigan, Michigan Handbook Series No. 3. 229 pp.[facsimile reprint 2007 by Center for North American Herpetology, Facsimile Reprint Series, No.7].
Sheldon, A.B. 1991. Lizards they're not. Michigan Natural Resources Magazine 60(2):36–41.
Sheldon, A.B. 2006. Amphibians and Reptiles of the North Woods. Kollath and Stensaas, Duluth, MN. 148 pp.
Sheldon, A.B. 2021. Amphibians and Reptiles of Minnesota, Wisconsin, and Michigan.

Kollath and Stensaas, Duluth, MN. 201 pp.

Smith, W.H. 1879. Catalogue of the Reptilia and Amphibia of Michigan. Supplement to Science News, I pp. I–VIII.

Switzenberg, D.F. 1959. Michigan Snakes. Michigan Department of Conservation, Lansing. 6 pp.

Switzenberg, D.F. and Karl F. Lagler. 1952. Michigan Snakes and Turtles. Michigan Department of Conservation, Education Division, Lansing, MI. 9 pp.

Tekiela, S. 2004. Reptiles and Amphibians of Michigan Field Guide. Adventure Publications, Cambridge, MN. Xxiv + 145 pp. + CD.

Tekiela, S. 2014. Reptiles and Amphibians of Minnesota, Wisconsin and Michigan. Adventure Publications, Cambridge, MN. 216 pp.

Thompson, C. 1915. The reptiles and amphibians of Manistee County, Michigan. Occasional Papers of the Museum of Zoology, University of Michigan 18. 6 pp.

Minnesota

Blasus, R.E. 1997. Amphibian and Reptile Time Table for Minnesota. Minnesota Herpetological Society Occasional Paper No. 4. 30 pp.

Breckenridge, W.J. 1938. Minnesota lizards. Minneapolis Public Library Museum Nature Note 1:10–12.

Breckenridge, W.J. 1940. Reptiles and amphibians of Minnesota. Minneapolis Public Library Museum Nature Note 3:411–418

Breckenridge, W.J. 1941. Minnesota turtles. Conservation Volunteer 2(7):11–16.

Breckenridge, W.J. 1942. Frogs and toads of Minnesota. Conservation Volunteer 5(27):32–36.

Breckenridge, W.J. 1942. Minnesota's non–poisonous snakes. Conservation Volunteer 4(21):10–15.

Breckenridge, W.J. 1943. Do you recognize Minnesota's lizards? Conservation Volunteer 6(33):21–24.

Breckenridge, W. J. 1943. Those puzzling salamanders. Conservation Volunteer 6(31):9–12.

Breckenridge, W.J. 1943. Reptiles of Minnesota. Minnesota Department of Conservation, Conservation Bulletin No. 3. 24 pp.

Breckenridge, W.J. 1944. **Reptiles and Amphibians of Minnesota**. University of Minnesota Press, Minneapolis, MN. xiii + 202 pp. [reprinted undated and 1970].

Christoffel, R., J. Edwards, and B. Perry. 2010. Snakes and Lizards of Minnesota. Minnesota Nongame Program, St. Paul, MN. 68 pp.

Ernst, C.H. 1973. The distribution of the turtles of Minnesota. Journal of Herpetology 7:42–47.

Gerholdt, J. 1999. Frogs and Toads of Minnesota. Bell Museum of Natural History Leaflet No. 11. 12 pp.

Karns, D.R. 1986. Field Herpetology: Methods for the Study of Amphibians and Reptiles in Minnesota. Bell Museum of Natural History Occasional Paper No. 18. 88 pp.

Lang, J.W. and D.R. Karns. 1988. Amphibians and Reptiles pp. 323–349 *in* B. Coffin and L. Pfannmuller, (eds.) Minnesota's Endangered Flora and Fauna. University of Minnesota Press, Minneapolis, MN. xv + 473 pp.

Moriarty, J.J. 1987. Distribution Maps for Reptiles and Amphibians of Minnesota. Minnesota Herpetological Society, Minneapolis, MN. 50 pp.

Moriarty, J.J. 1989. Turtles of Minnesota. Bell Museum of Natural History Leaflet No. 9. 4 pp.

Moriarty, J. J. 1990. Meet Minnesota's salamanders. Minnesota Volunteer 53 (2):28–31.

Moriarty, J.J. 1995. Leaping leopards and other frogs, toads, and treefrogs. Minnesota Volunteer 58(3):20–33.

Moriarty, J.J. 1997. Have home, will travel: Turtles of Minnesota. Minnesota Volunteer. 60(4):38–47.

Moriarty, J.J. 2004. Bibliography of Minnesota Herpetology Through 2003. Minnesota Herpetological Society Occasional Paper Number 6. 46 pp.

Moriarty, J.J. 1998. Status of Amphibians in Minnesota pp 166–169 *in* M.J. Lannoo, (ed.) Status and Conservation of Midwestern Amphibians. University of Iowa Press, Iowa City, IA. xviii + 507 pp.

Moriarty, J.J. and D.G. Jones. 1988. An Annotated Bibliography of Minnesota Herpetology 1900–1985. Bell Museum of Natural History, Minneapolis, MN. 36 pp.

Moriarty, J.J. and D. Jones. 1997. Minnesota's Amphibians and Reptiles: Their Conservation and Status. Proceedings of a Symposium. Serpent's Tale Natural History Books, Lanesboro, MN. 75 pp.

Moriarty, J.J., and C.D. Hall. 2014. **Amphibians and Reptiles in Minnesota**. University Minnesota Press, Minneapolis, MN. xii + 370 pp.

Oldfield, B. and J.J. Moriarty. 1994. Amphibians and Reptiles Native to Minnesota. University of Minnesota Press. Minneapolis, MN. xii + 237 pp.

Parmelee, J.R., M.G. Knutson, and J.E. Lyon. 2002. A Field Guide to Amphibian Larvae and Eggs of Minnesota, Wisconsin, and Iowa. U.S. Geological Survey Information and Technology Report 2002–0004. 38 pp.

Perry, P.S. and M.H. Dexter. 1989. Snakes and Lizards of Minnesota. Minnesota Nongame Wildlife Program, St. Paul, MN. 25 pp.

Sheldon, A.B. 2006. Amphibians and Reptiles of the North Woods. Kollath and Stensaas, Duluth, MN. 148 pp.

Sheldon, A.B. 2021. Amphibians and Reptiles of Minnesota, Wisconsin, and Michigan. Kollath and Stensaas, Duluth, MN. 201 pp.

Swanson, G.1935. A preliminary list of Minnesota amphibians. Copeia 1935:152– 154.

Tekiela, S. 2003. Reptiles and Amphibians of Minnesota. Adventure Publications, Cambridge, MN. 140 pp. + CD.

Tekiela, S. 2014. Reptiles and Amphibians of Minnesota, Wisconsin and Michigan. Adventure Publications, Cambridge, MN. 216 pp.

Mississippi

Allen, M.J. 1932. A survey of the amphibians and reptiles of Harrison County, Mississippi. American Museum Novitiates 542:1–20.

Cliburn, J. 1965. A Key to the Amphibians and Reptiles of Mississippi. Mississippi Fish and Game Commission, Jackson, MS. iv + 74 pp. [2nd edition undated ([iv] + 70 pp.), 3rd edition 1970 ([iv] + 63 pp.), 4th edition 1976 ([iv] + 72 pp.)].

Cliburn, J.W. and C.G. Jackson, Jr. 1975. Rare and Endangered Amphibians and Reptiles in Mississippi. Pp. 12–20 *in* A Preliminary List of the Rare and Threatened Vertebrates in Mississippi. Game and Fish Commission, Jackson, MS. i + 29 pp.

Cook, F.A. 1942. Alligator and Lizards of Mississippi. Mississippi Game and Fish Commission, Jackson, MS. [2] + v + [1] + 20 pp. [Reprinted 1943, 1957 ([1] + v + 20), 1966 ([1] + v + 20)].

Cook, F.A. 1943. Snakes in Mississippi. Survey Bulletin, Mississippi Game and Fish Commission, Jackson, MS. [2] + ii + 73 pp.

Cook, F. A. 1954. **Snakes of Mississippi**. Mississippi Game and Fish Commission, Public Relations Department, Jackson, MS. [2] + ii + 40 pp. [revised 1962, [2] + ii + 45 pp.].

Cook, F.A. 1957. Salamanders of Mississippi. Mississippi Game and Fish Commission, Jackson, MS. [2] + ii + 28 pp. [undated reprint].

Gandy, B.E. 1966. A Preliminary Checklist of the Vertebrates of Mississippi. Mississippi State Wildlife Museum. Jackson. 34 pp.

Keiser, E.D. 1982. The poisonous snakes of Mississippi with suggestions for the emergency treatment of their bite. Mississippi Outdoors 1982 (4):1a–16a.

Keiser, E.D. 2001. **Turtles of the University of Mississippi Field Station**. University of Mississippi Field Station, Abbeville, MS. 20 pp.

Keiser, E.D. 2008. Frogs of the University of Mississippi Field Station. University of Mississippi Field Station, Abbeville, MS. 59 pp.

Keiser, E.D. 1999. Salamanders of the University of Mississippi Field Station. University of Mississippi Field Station, Abbeville, MS. 20 pp.

Keiser, E.D. 2010. Lizards of the University of Mississippi Field Station. University of Mississippi Field Station, Abbeville, MS. 59 pp.

Keiser, D. 2010. Snakes of the University of Mississippi Field Station. University of Mississippi Field Station. Abbeville, MS. 97 pp.

Lohoefener, R. and R. Altig. 1983. Mississippi Herpetology. Mississippi State University Research Center Bulletin 1. Vi + 66 pp.

Smith, P.W. and J.C. List. 1955. Notes on Mississippi amphibians and reptiles. American Midland Naturalist 53:115–125.

Missouri

Anderson, P. 1942. Amphibians and reptiles of Jackson County, Missouri. Bulletin of Chicago Academy of Science. 6:203–228.

Anderson, P. 1965. **The Reptiles of Missouri**. University of Missouri Press, Columbia, MO. xxiii + 330 pp.

Boyer, D.A. and Heinze, A.A. 1934. An annotated list of the amphibians and reptiles of Jefferson County, Missouri. Transactions of the Academy of Science of St. Louis 28(4):185–200, 1 pl.

Briggler, J.T. 2023. A Guide to Missouri's Lizards. Missouri Department of Conservation, Jefferson City, MO. 36 pp.

Briggler, J.T. and T.R. Johnson. 2008. Missouri's Toads and Frogs. Missouri Conservation Commission, Jefferson City, MO. 32 pp.

Briggler, J.T. and T.R. Johnson. 2015. Missouri's Turtles. Missouri Conservation Commission, Jefferson City, MO. 16 pp.

Briggler, J.T. and T.R. Johnson. 2017. A Guide to Missouri's Snakes. Missouri Conservation Commission, Jefferson City. MO. 59 pp. + 8 pl.

Briggler, J.T. and T.R. Johnson. 2021. **Amphibians and Reptiles of Missouri**. Missouri Conservation Commission, Jefferson City. MO. 514 pp.

Briggler, J.T. and T.R. Johnson. 2022. A Guide to Missouri's Toads and Frogs. Missouri Department of Conservation, Jefferson City, MO. 37 pp.

Briggler, J.T. and T.R. Johnson. 2023. A Guide to Missouri's Turtles. Missouri Department of Conservation, Jefferson City, MO. 32 pp.

Briggler, J.T. and T.R. Johnson. 2023. A Guide to Missouri's Snakes. Missouri Department of Conservation Jefferson City, MO. 59 pp.

Casey, J.T. and S. Cox. 2021. Amphibians and Reptiles of Northwest Missouri. Pony Express Amphibian and Reptile Society, St. Joseph, MO. 60 pp.

Hurter, J. 1893. Catalogue of reptiles and batrachians found in the vicinity of St. Louis, Missouri. Transactions of St. Louis Academy of Science 6:251–261.

Hurter, J. 1911. Herpetology of Missouri. Transactions of the Academy of Science of St. Louis 20:59–274 + 7 pls.

Johnson, T.R. 1974. A preliminary checklist of the salamanders of Missouri. St. Louis Herpetological Society Newsletter 1(3):8–9.

Johnson, T.R. 1974. A preliminary checklist of Missouri toads and frogs. St. Louis Herpetological Society Newsletter 1(2):6–7.

Johnson, T.R. 1977. The Amphibians of Missouri. University of Kansas Museum of Natural History Public Education Series No. 6. ix + 134 pp.

Johnson, T.R. 1979. Missouri's venomous snakes. Missouri Conservationist 40(6):4–7.

Johnson, T.R. 1980. Snakes of Missouri. Missouri Conservationist 41(4):11–22.

Johnson, T.R. 1982. Missouri's turtles. Missouri Conservationist 42(6):11–22.

Johnson, T.R. 1982. Missouri's toads and frogs. Missouri Conservationist 43(6):11–22

Johnson, T.R. 1987. The Amphibians and Reptiles of Missouri. Missouri Conservation Commission, Jefferson City, MO. Xi + 368 pp.

Johnson, T. R. 1997. The Lizards of Missouri. Conservation Commission of the State of Missouri, Jefferson City, Missouri. Unpaginated [8 pp.].

Johnson, T.R. 2000. The Amphibians and Reptiles of Missouri. Missouri Department of Conservation, Jefferson City, MO. 400 pp.

Johnson, T.R. and R.N. Bader 1974. Annotated checklist of Missouri amphibians and reptiles. St. Louis Herpetologist Society. Special Issue 1:1–16.

Miller, J. 2010. Show–Me Herps. Missouri Conservation Commission, Jefferson City MO. 152 pp.

Schroeder, E.E. and T.S. Baskett. 1965. Frogs and Toads of Missouri. Missouri Conservationist. 11 pp. [3 parts: 26(1):15–16, and back cover, 26(3):16, and back cover, 26(4):15–16, and back cover].

Schwartz, C.W. 1950. Snakes and Facts About Them. Missouri Conservation Commission. 8 pp.

Wiley, J.R. 1968. Guide to the amphibians of Missouri. Missouri Speleologist 10(4):132–172.

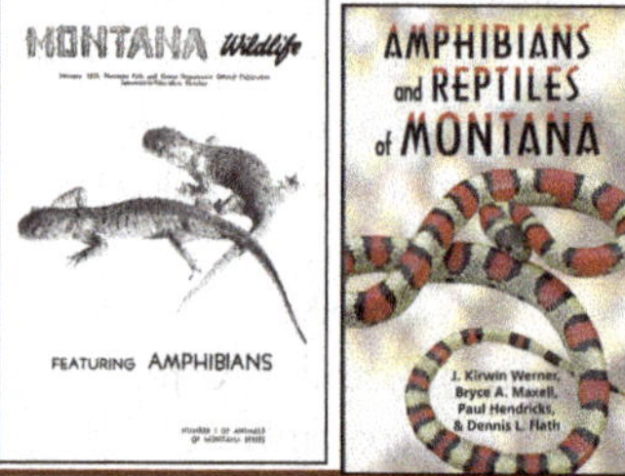

Montana

Black, J.H. 1967. Toads of Montana. Montana Wildlife, January 1967(Spring):22–28.

Black, J.H. 1970. **Amphibians of Montana**. Montana Wildlife: Animals of Montana Series, 1970(1): 1–32.

Black, J.H. 1970. Turtles of Montana: Montana Wildlife: No. 2 of Animals of Montana Series. November 1970: 26–31.

Coues, E. and H.C. Yarrow. 1878. Notes on the herpetology of Dakota and Montana. Bulletin of the United States Geological Survey 4:259–291. [reprinted 1978 *in* Adler, K. (ed.), Herpetological Explorations of the Great American West, Volume 1. Arno Press, New York, NY].

Davis, C.V. and S.E. Weeks. 1963. Montana Snakes. Montana Wildlife. 11 pp.

Koch, E.D. and C.R. Peterson. 1995. Amphibians and Reptiles of Yellowstone and Grand Teton National Parks. University of Utah Press, Salt Lake City, UT. xviii + 188 pp.

Maxell, B.A. 2000. Management of Montana's Amphibians. U.S. Forest Service, Northern Regional Office Region 1, Missoula, MT. 161 pp.

Maxell, B.A., J.K. Werner, P. Hendricks, and D.L. Flath. 2003. Herpetology in Montana. Society for Northwestern Vertebrate Biology, Olympia, WA. Viii + 138 pp.

Mosiman, J.E. and G.B. Rabb. 1952. The herpetology of the Tiber Reservoir area, Montana. Copeia 1952:23–27.

Reichel, J. and D. Flath. 1995. Identification of Montana's amphibians and reptiles.

Montana Outdoors 26(3):15–34.

Thompson, L.S. 1982. Distribution of Montana Amphibians, Reptiles, and Mammals. Montana Audubon Council, Helena, MT. 24 pp.

Turner, F.B. 1955. Reptiles and Amphibians of Yellowstone Park. Yellowstone National Park, Yellowstone Interpretive Series No. 5. Vi + 40 pp.

Werner, J.K., B.A. Maxell, P. Hendricks, and D.L. Flath. 2004. **Amphibians and Reptiles of Montana**. Mountain Press, Missoula, MT. xii + 262 pp.

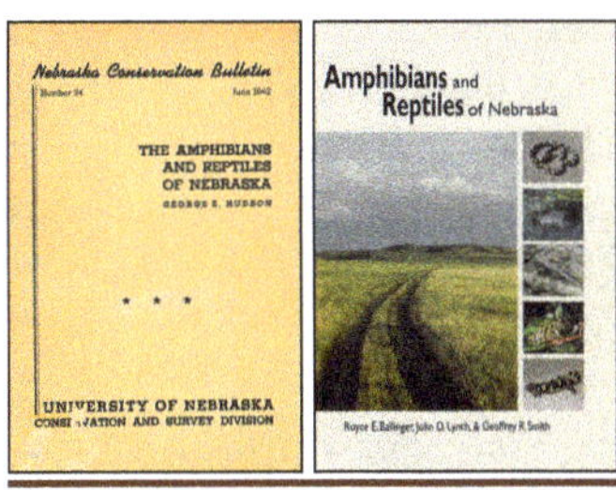

Nebraska

Ballinger, R.E., J.D. Lynch, and P.H. Cole. 1979. Distribution and natural history of amphibians and reptiles in western Nebraska with ecological notes on the herptiles of Arapaho Prairie. Prairie Naturalist 11:65–74.

Ballinger, R.E., J.D. Lynch, and G.R. Smith. 2010. **Amphibians and Reptiles of Nebraska**. Rusty Lizard Press, Oro Valley, AZ. 400 pp.

Benedict, R. 1996. Snappers, softshells, and stinkpots: the turtles of Nebraska. Museum Notes, University of Nebraska State Museum 96. 4 pp. text, fold out poster.

Fogell, D.D. 2010. Amphibians and Reptiles of Nebraska. Institute of Agriculture and Natural Resources, Lincoln, NE. 155 pp.

Hudson, G.E. 1942. **The Amphibians and Reptiles of Nebraska**. University of Nebraska, Nebraska Conservation Bulletin No. 24. 146 pp. [reprinted 1958].

Hudson, G.E. and Davis, D. 1940. Facts concerning the snakes of Nebraska. University of Nebraska, Department of Zoology. 4 pp. [revised in 1941 and possibly later].

Lynch, J.D. 1985. Annotated checklist of the amphibians and reptiles of Nebraska. Transactions of the Nebraska Academy of Science 13: 33–57.

Taylor, W.E. 1892. The Ophidia of Nebraska. Annual Report of the Nebraska State Board of Agriculture 1891:310–357.

Taylor, W.E. 1892. Catalogue of the snakes of Nebraska with notes on their habits and distribution. The American Naturalist 26:742–752.

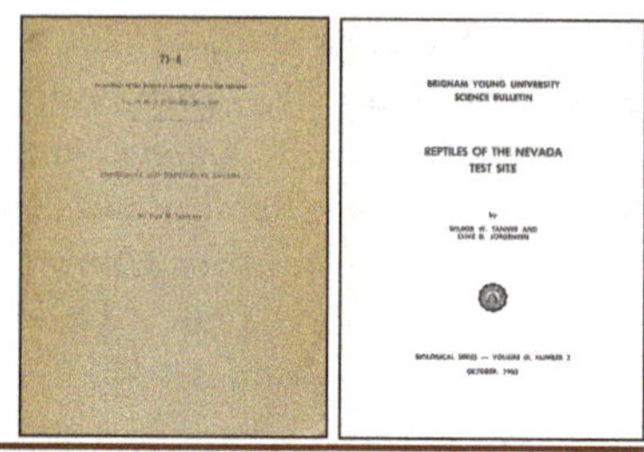

Nevada

Banta, B.H. 1965. An annotated chronological bibliography of the herpetology of the State of Nevada. Wasmann Journal of Biology 23:1–224.

Banta, B.H. 1965. A distributional check list of the Recent reptiles inhabiting the State of Nevada. Biological Society of Nevada Occasional Paper 5. 8 pp.

Banta, B.H. 1965. A distributional check list of the Recent amphibians inhabiting the State of Nevada. Biological Society of Nevada Occasional Paper 7. 4 pp.

Cowles, R.B. and C.M. Bogert. 1936. The herpetology of the Boulder Dam region. Herpetologica 1:33–42.

Linsdale, J.M. 1940. **Amphibians and reptiles of Nevada**. Proceedings of the American Academy of Arts and Science 73:197–257.

Richardson, C.H. 1915. Reptiles of northwestern Nevada and adjacent territory. Proceedings of the United States National Museum 48:403–435.

Ruthven, A.G. and H.T. Gaige. 1915. The reptiles and amphibians collected in northeastern Nevada by the Walker–Newcomb Expedition of the University of Michigan. Occasional Papers of the Museum of Zoology University of Michigan 8. 33 pp. + 5 pls.

Tanner, W.W. and C.D. Jorgenson. 1963. **Reptiles of the Nevada Test Site**. Brigham Young University Science Bulletin 3 (3):1–31.

Van Denburgh, J. and J.R. Slevin. 1921. A list of the amphibians and reptiles of Nevada, with notes on the species in the collection of the Academy. Proceedings of the California Academy of Sciences, 4th Series 11:27–38.

New England

Babcock, H.L. 1919. The Turtles of New England. Memoirs of the Boston Society of Natural History 8:325–431, pls. 17–32. [facsimile reprints in 1971 by Dover Publications, New York, NY and in 2007 by Center for North American Herpetology, Facsimile Reprint Series, No. 9].

Babcock, H.L. 1929. The Snakes of New England. Boston Society of Natural History Natural History Guide No.1. 30 pp. + 4 pls.

Babcock, H.L. 1930. New England lizard records. Bulletin of the Boston Society of

Natural History 57:9–12.

Babcock. H.L. 1938. Field Guide to New England Turtles. Boston Society of Natural History Natural History Guide No.2. 56 pp. + 9 pls.

DeGraff, R.M. and D.D. Rudis. 1983. **Amphibians and Reptiles of New England: Habitats and Natural History**. University of Massachusetts Press, Amherst, MA. vii + 85 pp. [hardcover and softcover versions].

Dunn, E.R. 1930. New England salamanders. Bulletin of the Boston Society of Natural History 57:23–24.

Epple, A.O. 1983. The Amphibians of New England. Down East Books. Camden, ME. xvi +138 pp.

Henshaw, S. 1904. Fauna of New England: List of the Reptilia. Occasional Papers of The Boston Society of Natural History 7(1):1–13.

Henshaw, S. 1904. Fauna of New England: List of the Batrachia. Occasional Papers of the Boston Society of Natural History 7(2):1–10.

Himmelman, J. 2006. Discovering Amphibians: Frogs and Salamanders of the Northeast. Down East Books, Camden, ME. 208 pp.

Knobel, E. 1896. **Turtles, Snakes, and Frogs and Other Reptiles and Amphibians of New England and the North**. Bradlee Whidden, Boston, MA. 47 pp.

Krulikowski, L. 2005. Snakes of New England. Privately published, Old Lyme, CT. 307 pp.

MacCoy, C.V. 1931. Key for identification of New England amphibians and reptiles. Bulletin of the Boston Society of Natural History 59:25–33.

Massman, W.H. n.d. [1939 or later]. The Snakes of New England. A Field Guide. W.H. Massmann, Westport, CT. 24 pp.

Tyning, T. F. 1993. New England's turtles. Sanctuary: The Journal of the Massachusetts Audubon Society 32(5):5-7.

Tyning, T. 1995. A guide to New England frogs. Sanctuary: The Journal of the Massachusetts Audubon Society 34:16–17.

New Hampshire

Allen, G.M. 1899. Notes on the reptiles and amphibians of Intervale, New Hampshire. Proceedings of the Boston Society of Natural History. 29:63–75.

Huse, W.H. 1900. The Testudinata of New Hampshire. Journal of Proceedings of the Manchester Institute of Arts and Sciences 2:47–51.

Huse, W.H. 1901. **The Ophidia of New Hampshire**. Journal of Proceedings of the Manchester Institute of Arts and Sciences 3:115–121.

Oliver, J. A. and J.R. Bailey 1939. Amphibians and reptiles of New Hampshire, exclusive of marine forms. Biological Survey of Connecticut Watershed Report 4:195–217.

Taylor, J. 1993. **The Amphibians and Reptiles of New Hampshire**. New Hampshire Fish and Game Department, Concord, NH. 71 pp.

Taylor, J. 1997. The Status of Turtles in New Hampshire. pp.4–5 *in* T.F. Tyning, (ed.) Status and Conservation of Turtles of the Northeastern United States. Serpent's Tale Natural History Books, Lanesboro, MN. ix + 53 pp.

New Jersey

Abbot, C.C. 1869. Catalogues of Vertebrate Animals of New Jersey. Appendix E *in* New Jersey Geological Survey. Geology of New Jersey. Newark, NJ, xxiv + 899 pp. [amphibians and reptiles on pages 799–805].

Fowler, H.W. 1907. Amphibians and reptiles of New Jersey. Report of the New Jersey State Museum 1906: 23–250, 402–408 + 69 pl.

Fowler, H.W. 1908. A supplementary account of New Jersey amphibians and reptiles. Report of the New Jersey State Museum. 1907:190–202.

Lentz, A. N., P. G. Pearson, and J. R. Westman. 1962. A Checklist of Snakes, Lizards, Turtles, Frogs, and Salamanders in New Jersey. Rutgers State University Extension Service, College of Agriculture, New Brunswick, N. J. Leaflet 322. 8 pp.

McLain, P. 1979. The snakes of New Jersey. New Jersey Outdoors 6(3):6–7,25,30–32. [also published with McLain, P. 1979. More on the snakes of New Jersey. New Jersey Outdoors 6(4):12–13,28–29, 32 in a single 10 pp. undated reprint].

Nelson, J. 1890. Descriptive Catalogue of the Vertebrates of New Jersey. pp. 489–824 *in* Final Report of the State Geologist, Volume II. Geological Survey of New Jersey, Trenton, NJ. [amphibians and reptiles on pages 637–657].

Penn, N. 2020. NJ State Turtle Guide. privately published. 48 pp.

Schwartz, V. and D.M. Golden. 2002. **Field Guide to Reptiles and Amphibians of New Jersey**. New Jersey Fish and Wildlife, Woodbine, NJ. 89 pp. + CD.

Stiles, E.D. 1978. Vertebrates of New Jersey. Privately Published, Somerset, NJ. 148 pp. [amphibians and reptiles on pp. 34–63].

Trapido, H. 1937. **A Guide to the Snakes of New Jersey**. Newark Museum, Newark, NJ. 60 pp.

New Mexico

Bartlett, R.D. and P.P. Bartlett. 2013. **New Mexico's Reptiles and Amphibians A Field Guide**. University of New Mexico Press, Albuquerque, NM. 228 pp.

Degenhardt, W.G. and J.L. Christiansen. 1974. Distribution and habits of turtles in New Mexico. Southwest Naturalist 19:21–46.

Degenhardt, W.G., C.W. Painter, and A.H. Price. 1996. **The Amphibians and Reptiles of New Mexico**. University of New Mexico Press. Albuquerque, NM. xix + 504 pp. [reprinted 2005 in softbound edition].

Foxx, T.S., T.K. Haarmann, and D.C. Keller. 1999. Amphibians and Reptiles of Los Alamos County, New Mexico. Los Alamos National Laboratory, Los Alamos, NM. vi + 84 pp.

Gehlbach, F.R. 1965. Herpetology of the Zuni Mountains region of northwestern New Mexico. Proceedings of the United States National Museum 116:243–332.

Herrick, C.L., J. Terry, and H.N. Herrick, Jr. 1899. Notes on a collection of lizards from New Mexico. Bulletin of the Scientific Laboratories of Denison University 11:117–148 + pls. 14–24.

Hubbard, J.P., M.C. Conway, Howard Campbell, Greg Schmitt, and Mike D. Hatch. 1978. Handbook of Species Endangered in New Mexico. New Mexico Department of Game and Fish. Santa Fe, NM. vi + 202 pp.

Lee, L. 1961. Rattlesnakes of New Mexico … and where to find them. New Mexico Wildlife 6(2):12–13.

Livo, L.J. 1995. Identification Guide to Montane Amphibians of Southern Rocky Mountains. Colorado Division of Wildlife, Denver, CO. 25 pp.

Mosauer, W. 1932. The amphibians and reptiles of the Guadalupe Mountains of New Mexico and Texas. Occasional Papers of the Museum of Zoology University of Michigan.246. 18 pp. + 1 pl.

Painter, C.W. and J.N. Stuart. 2015. Herpetofauna of New Mexico / Herpetofauna de Nuevo Mexico. Pp. 164–180, 381–398 *in* Lemos-Espinal, J.A. (ed.), Amphibians and Reptiles of the US-Mexico Border States / Anfibios y Reptiles de los Estados de la Fontera México–Estados Unidos. Texas A&M University Press, College Station, TX. [checklist, plates and index collective for entire book].

Painter, C.W., J.N. Stuart, J.T. Giermakowski, and L.J.S. Pierce. 2017. Checklist of theamphibians and reptiles of New Mexico, USA, with notes on taxonomy, status, and distribution. Western Wildlife 4:29–60.

Prival, D. and M. Goode. 2011. Chihuahuan Desert National Parks Reptile and Amphibian Inventory. National Park Service – Natural Resources Report, Fort Collins, CO. 90 pp.

Ruthven, A.G. 1907. A collection of reptiles and amphibians from southern New Mexico and Arizona. Bulletin of the American Museum of Natural History 23:483–604.

Stuart, J.N. 2003. A Supplemental Bibliography of Herpetology in New Mexico. University of New Mexico, Albuquerque, NM. 60 + A1–A9 + B1 pp. [revised 2004 (59 + A1–A9 + B1 pp.), revised 2005 (66 + A1–A9 pp.); [PDF only - https://repositories.lib.utexas.edu/items/2a593dff-7acb-4401-82cd-b74391cc2059].

Tanner, D.L. 1975. Lizards of the New Mexican Llano Estacado and its adjacent river valleys. Studies in Natural Sciences, Eastern New Mexico University 2(2):1–39.

Van Denburgh, J. 1924. Notes on the herpetology of New Mexico, with a list of species from that state. Proceedings of the California Academy of Sciences, 4th Series 13: 189–230.

Williamson, M.A., P.W. Hyder, J.S. Applegarth. 1994. Snakes, Lizards, Turtles, Frogs, Toads, and Salamanders of New Mexico. Sunstone Press, Santa Fe, NM. iv + 176 pp.

New York

Anonymous. 1968. The Amphibia of Long Island. Sanctuary, Summer 1968: 1–19.

Baird, S.F. 1854. **On the Serpents of New York with a notice of a species not hitherto included in the fauna of the state.** Annual Report of the New York State Cabinet of Natural History 7:95–124 + 2 pls.

Bishop, S.C. 1927. The Amphibians and Reptiles of Allegheny State Park, New York State Museum Handbook No. 3:5–134.

Bishop, S.C. 1941. The Salamanders of New York. New York State Museum Bulletin No. 324. 365 pp.

Block, S. 1985. Salamanders of New York. The Conservationist 39(5):42–47

Breisch, A. 1997. The Status and Management of Turtles in New York. pp. 11–14 in T.F. Tyning, (ed.) Status and Conservation of Turtles of the Northeastern United States. Serpent's Tale Natural History Books, Lanesboro, MN. ix + 53 pp.

Breisch, A. and J.L. Behler. 2002. Turtles of New York State. New York State Conservationist 57(1):15–18.

Breisch, A. and P.K. Ducey. 2003. Lake, pond & stream salamanders of New York State. New York State Conservationist 57(5):15–18.

Breisch, A. and P.K. Ducey. 2003. Woodland & vernal pool salamanders of New York State. New York State Conservationist 57(6):15–18.

Breisch, A. and J.P. Gibbs. 2002. Frogs & toads. New York State Conservationist 56(5)15–18.

Clausen, R.T. 1943. Amphibians and reptiles of Tioga County, New York. American Midland Naturalist 29:360–364.

Cook, R. P., D. K. Brotherton, and J. L. Behler. 2010. Inventory of Amphibians and Reptiles at Sagamore Hill National Historic Site. U.S. Department of the Interior, National Park Service, Natural Resource Program Center, Natural Resource Report NPS/NCBN/NRTR–2010/379. 57 pp.

Cook, R. P., D. K. Brotherton, and J. L. Behler. 2010. Inventory of Amphibians and Reptiles at Fire Island National Seashore. U.S. Department of the Interior, National Park Service, Natural Resource Program Center, Natural Resource Report NPS/NCBN/NRTR–2010/378. 101 pp.

De Kay, J. 1842. Zoology of New York, Part I: Reptiles and Fishes. State of New York, Albany, NY. 2 volumes, vii + 415 pp + 79 pl [reptile and amphibian section - (text) vii + 98 pp., (plates) 23 pls.].

Ditmars, R.L. 1896. The snakes found within fifty miles of New York City. Abstract of the Proceedings of the Linnean Society of New York 8:9–24.

Ditmars, R.L. 1905. The reptiles in the vicinity of New York City. American Museum Journal 5:93–140. [reprinted as Guide Leaflet No.19. i–v + 93–140].

Ditmars, R.L. 1905. The batrachians in the vicinity of New York City. American Museum Journal 5:161–206. [reprinted as Guide Leaflet No.20: 1–6 + 161–206].

Eckel, E.C. 1901. The snakes of New York State: An annotated check list. American Naturalist 35:151–155.

Eckel, E.C. and F.C. Paulmier. 1902. Catalogue of New York reptiles and batrachians. Bulletin of the New York State Museum 51:356–414.

Gibbs, J.P., A.R. Breisch, P.K. Ducey, G. Johnson, J.L. Behler, and R.C. Bothner. 2007. **The Amphibians and Reptiles of New York State**. Oxford University Press, New York, NY. 422 pp.

Lawrence, J.E. 1953. Snakes of the Catskill Mountains: A Guide to Their Recognition. privately published, Woodstock, NY. 36 pp. [reprinted by Outdoor Publications, Ithaca, NY, undated].

Mearns, E.A. 1898. A study of the vertebrate fauna of the Hudson Highlands, with observations on the Mollusca, Crustacea, Lepidoptera, and the flora of the region. Bulletin of the American Museum of Natural History 10:303–352. [amphibians and reptiles on pages 322–330].

Noble, G.K. 1927. Distributional List of the Reptiles and Amphibians of the New York City Region. American Museum of Natural History Guide Leaflet Series 69. 9 pp. [revised in 1929, 16 pp. (blank "Notes" pages unnumbered in original but numbered in revision)].

Overton, F. 1914. Long Island fauna and flora III. The frogs and toads. Museum of Brooklyn Institute of Arts and Science, Science Bulletin 2:21–40 + pls. 2–13.

Palmer, E.L. 1922. Amphibia and Reptilia. Cornell Rural School Leaflet 15(4): 303–364.

Palmer, E.L. 1947. Salamanders, toads and frogs. Cornell Rural School Leaflet 40(4):1–32.

Palmer, E.L. 1952. Reptiles. Cornell Rural School Leaflet 45(4):1–31.

Reilly, Jr., E.M. 1955. Snakes of New York. The New York State Conservationist 9(6):22–26.

Reilly, Jr., E.M. 1957. Salamanders and lizards of New York. The New York State Conservationist 11(6):23–28, 35.

Reilly, Jr, E.M. 1958. Turtles of New York. The New York State Conservationist 12(6):22–27.

Rockcastle, V.N. 1960. Reptiles. Cornell Science Leaflet 53. 34 pp.

Rockcastle, V.N. 1961. Amphibians. Cornell Science Leaflet 54. 32 pp.

Sherwood, W.L. 1895. The salamanders found in the vicinity of New York City, with notes upon extra-limital and allied species. Abstract of the Proceedings of the Linnean Society of New York No. 7:21–37.

Sherwood, W.L. 1898. The frogs and toads in the vicinity of New York City. Abstract of the Proceedings of the Linnean Society of New York 10:9–24.

Smith, E. 1899. The turtles and lizards of the vicinity of New York City. Abstract of the Proceedings of the Linnean Society of New York 11: 11–32.

Wright, A.H. 1914. Life-histories of the Anura of Ithaca, New York. Carnegie Institution of Washington, Publication. No. 197. 98 pp. + 13 pls.

Wright, A.H. 1955. Frogs and toads of New York. The New York State Conservationist 10(1):23–26.

North Carolina

Beane, J., A.L. Braswell, J.C. Mitchell, W.M. Palmer, and J.R. Harrison. 2010. **Amphibians and Reptiles of the Carolinas and Virginia**. University of North Carolina Press, Chapel Hill, NC. 274 pp.

Braswell, A.L. 1988. A survey of the amphibians and reptiles of Nags Head Woods Ecological Reserve. Association of Southeastern Biologists Bulletin 35:199–217.

Braswell, A.L., W.M. Palmer, and J.C. Beane. 2003. Venomous Snakes of North Carolina. North Carolina State Museum, Raleigh, NC. 31 pp. [reprinted 2004 and subsequently].

Breeder, C.M. 1923. A list of fishes, amphibians and reptiles collected in Ashe County, North Carolina. Zoologica (New York) 4(1):1–23.

Brimley, C.S. 1907. Artificial key to the species of snakes and lizards which are found in North Carolina. Journal of the Elisha Mitchell Science Society 23:141–149.

Brimley, C.S. 1915. List of the reptiles and amphibians of North Carolina. Journal of the Elisha Mitchell Science Society 30: 195–206.

Brimley, C.S. 1920. The turtles of North Carolina, with a key to the turtles of the Eastern United States. Journal of the Elisha Mitchell Science Society 36:62–71.

Brimley, C.S. 1926. Revised key and list of the amphibians and reptiles of North Carolina. Journal of the Elisha Mitchell Science Society 42:75–93.

Brimley, C.S. 1944. Amphibians and Reptiles of North Carolina. Carolina Biological Supply, Elon College, NC. 63 pp.

Brothers, D.R. 1965. An annotated list of the amphibians and reptiles of northeastern North Carolina. Journal of the Elisha Mitchell Science Society 81:119–124.

Brothers, D.R. 1992. An Introduction to the Snakes of the Dismal Swamp Region of North Carolina and Virginia. Edgewood PROBES, Inc., Boise, ID, viii + 139 pp. + 4 pls.

Brown, E.E. 1992. Notes of the amphibians and reptiles of the western Piedmont of North Carolina. Journal of the Elisha Mitchell Science Society 108:38–54.

Cooper, J.E., S.S. Robinson, and J.B. Funderburg (eds.). 1977. Endangered and Threatened Plants and Animals of North Carolina. North Carolina State

Museum of Natural History, Raleigh, NC. 444 pp.

DePoe, C.E., J.B. Funderburg, Jr., and T.L. Quay. 1961. The reptiles and amphibians of North Carolina: A preliminary checklist and bibliography. Journal of the Elisha Mitchell Science Society 77:125–136.

Dodd, Jr., K.C. 2004. The Amphibians of Great Smoky Mountains National Park. University of Tennessee Press, Knoxville, TN. 283 pp.

Dorcas, M.E. 2004. A Guide to the Snakes of North Carolina. Davidson College, Charlotte, NC. 40 pp.

Dorcas, M.E., S.J. Price, J.C. Beane, and S.C. Owen. 2007. The Frogs and Toads of North Carolina Field Guide and Recorded Calls. North Carolina Wildlife Resources, Raleigh, NC. 80 pp + CD.

Dunn, E.R. 1917. Reptiles and amphibians of the North Carolina mountains, with especial references to salamanders. Bulletin of American Museum of Natural History 37:593–634.

Funderburg, J.B. 1955. The amphibians of New Hanover County, North Carolina. Journal of the Elisha Mitchell Science Society 71:19–28.

Huheey, J.E. and A. Stupka. 1967. Amphibians and Reptiles of Great Smoky Mountains National Park. University of Tennessee Press, Knoxville, TN. x + 98 pp.

King, W. 1939. A survey of the herpetology of the Great Smoky Mountains National Park. American Midland Naturalist 21:531–582.

Martof, B.S., W.M. Palmer, J.R. Bailey, and J.R. Harrison. 1980. Amphibians and Reptiles of the Carolinas and Virginia. University of North Carolina Press, Chapel Hill, NC. [viii] + 264 pp. [undated softback reprint].

Palmer, W.M. 1964. The Snakes of North Carolina. North Carolina State Museum, Information Circular 64/3. 6 pp.

Palmer, W.M. 1974. Poisonous Snakes of North Carolina. State Museum of Natural Sciences, Raleigh, NC. 22 pp. [reprinted 1978, revised 1983, reprinted 1986, 5th printing 1990].

Palmer, W.M. and A. L. Braswell. 1995. **Reptiles of North Carolina**. Univ. of North Carolina Press, Chapel Hill, NC. xiii + 412 pp.

Tilley, S.G. and J.E. Huheey. 2001. Reptiles and Amphibians of the Smokies. Great Smoky Mountains Natural History Association, Gatlinburg, TN. 143 pp.

Travis, J. 1981. A key to the tadpoles of North Carolina. Brimleyana 6:119–127.

Weller, W.H. 1931. A preliminary list of the salamanders of the Great Smoky Mountains of North Carolina and Tennessee. Proceedings of the Junior Society of Natural Sciences 2:21–32. [reprinted 1965 by the Ohio Herpetological Society, pp. 25–36 *in* Herpetological Papers from the Proceedings of the Junior Society of Natural Sciences (1930–1932)].

North Dakota

Coues, E. and H.C. Yarrow. 1878. Notes on the herpetology of Dakota and Montana. Bulletin of the United States Geological Survey 4:259–291. [reprinted 1978 *in* Adler, K. (ed.), Herpetological Explorations of the Great American West, Volume 1. Arno Press, New York, NY].

Hoberg, T. and C. Gause. 1992. Reptiles and amphibians of North Dakota. North Dakota. Outdoors 55(1):7–20.

Johnson, S. 2015. **Reptiles and Amphibians of North Dakota**. North Dakota Game and Fish Department, Bismarck, ND 57 pp.

Wheeler, G.C. 1947. The amphibians and reptiles of North Dakota. American Midland Naturalist. 38 :162–190.

Wheeler, G.C. 1954. The Amphibians and Reptiles of the North Dakota Badlands. Roosevelt Nature and History Association, Medora, ND. 56 pp.

Wheeler, G.C., and J. Wheeler. 1966. **The Amphibians and Reptiles of North Dakota**. University of North Dakota Press, Grand Forks, ND. vii + 104 pp.

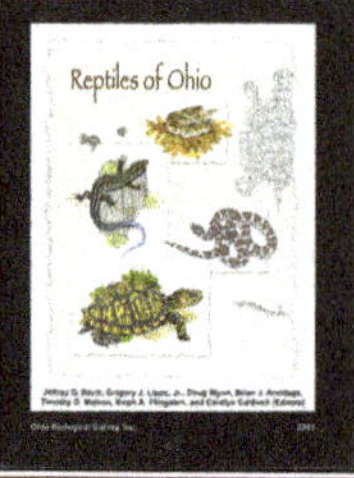

Ohio

Blem, C.R. 1972. An annotated list of the amphibians and reptiles of Hardin County, Ohio. Ohio Journal of Science 72:91–96.

Collins, J.D., John P. Mathews, and Joseph T. Collins. 2011. A Pocket Guide to Ohio Snakes. Center for North American Herpetology, Lawrence, KS. 57 pp.

Collins, J.T. and G.T. McDuffie. 1969. Salamanders of the Cincinnati region. The Explorer 11(1):21–24.

Collins, J.T. and G.T. McDuffie. 1970. Frogs and toads of the Cincinnati area. The Explorer 12(2):12–15.

Collins, J.T. and G.T. McDuffie. 1971. Turtles and lizards of Cincinnati region. The Explorer 13(3):26–29.

Collins, J.T. and G.T. McDuffie. 1972. Snakes of the Cincinnati region. The Explorer 14(1):24–28.

Conant, R. 1938. The reptiles of Ohio. American Midland Naturalist 20:1–200.

Conant, R. 1951. The Reptiles of Ohio. Second Edition. University of Notre Dame Press,

Notre Dame, IN. 284 pp.

Condit, J.M. 1973. Ohio Salamanders. Natural History Information Service, Ohio Historical Society 3(3):1–8.

Davis, J.G., G.J. Lipps, Jr., D. Wynn, B.J. Armitage, T.O. Matson, R.A. Pfingsten, and C. Caldwell (eds.). 2021. **Reptiles of Ohio**. Ohio Biological Survey, Columbus, OH. 2 vol., xx + 1112 pp.

Davis, J.G. and Scott A. Menze. 2000. Ohio Frog and Toad Atlas. Ohio Biological Survey, Columbus, OH. 23 pp.

Davis, J.G. and S.A. Menze. 2002. In Ohio's Backyards: Frogs and Toads. Ohio Biological Survey, Columbus, OH. x + 141 pp.

Denny, G. 1969. Reptiles of Lake County. Project P. L. E. A. S. E., Willoughby, OH. 85 pp.

Denny, G.L. 1978. Ohio's Amphibians. Ohio Department of Natural Resources, Columbus, OH. 29 pp.

Denny, G.L. n.d. Ohio's Reptiles. Ohio Department of Natural Resources, Columbus, OH. 25 pp.

Denny, G. 2006. Ohio's Amphibians. Ohio Division of Wildlife, Columbus, OH. 48 pp.

Denny, G. 2007. Ohio's Reptiles. Ohio Division of Wildlife, Columbus, OH. 36 pp

Dury, R. 1934. Snakes of the Cincinnati region. Zoological Society of Cincinnati Quarterly 1(1):7–12.

Gehlbach, F.R. 1960. Comments on the study of Ohio salamanders with key to their identification. Journal of Ohio Herpetological Society 2(3):10–15.

Kirtland, J.P. 1838. Report on the Zoology of Ohio. Second Annual Report of the Geological Survey of Ohio pp. 159–200. [amphibians and reptiles on pp. 167–168, 188–190; reprinted 1961 by the Ohio Herpetological Society].

Kraus, F. and G. Schuett. 1982. A Herpetofaunal Survey of the Coastal Zone of Northwest Ohio. Kirtlandia 36:21-54.

Langlois, T.H. 1964. Amphibians and reptiles of the Erie Islands. Ohio Journal of Science 64:11–25.

Matson, T.O., R.L. Muehlheim, and J.C. Spetz 2004. Survey of Fishes, Amphibians, and Reptiles of the Conneaut Creek Drainage System, Ashtabula County, Ohio. Kirtlandia, Cleveland, OH. 32 pp.

Morse, M. 1904. Batrachians and reptiles of Ohio. Proceedings of the Ohio State Academy of Sciences 4:91–144. Special Paper No. 9.

Ohio Department of Natural Resources, Division of Wildlife. 2008. Amphibians of Ohio. Ohio Division of Wildlife, Columbus, OH. 47 pp.

Ohio Department of Natural Resources, Division of Wildlife. 2018. Reptiles of Ohio Field Guide. Publication 5354 (0118). Ohio Division of Wildlife, Columbus, OH. 55 pp.

Ohio Department of Natural Resources, Division of Wildlife. 2019. Amphibians of Ohio Field Guide. Publication 5348(0119). Division of Wildlife, Columbus, OH. 46 pp.

Pfingsten, R.A. and F.L. Downs. 1989. **Salamanders of Ohio**. Bulletin of the Ohio Biological Survey. 7(2):1–315, 29 pls.

Pfingsten, R.A. 1998. Distribution of Ohio amphibians. pp. 221–255 *in* Lannoo, M.J. (ed.) Status and Conservation of Midwestern Amphibians. University of Iowa Press, Iowa City, IA. xviii + 507 pp.

Pfingsten, R.A. and T.O. Matson. 2003. Ohio Salamander Atlas. Ohio Biological Survey,

Columbus, OH. 32 pp.

Pfingsten, R.A., J.G. Davis, T.O. Matson, G.J. Lipps, D. Wynn, and B.J. Armitage (eds.). 2013. Amphibians of Ohio. Ohio Biological Survey, Columbus, OH. xiv + 899 pp.

Seibert, H.C. and R.S. Brandon. 1960. The salamanders of southeastern Ohio. Ohio Journal of Science. 60:291–303.

Smith, W.H. 1882. Report on the reptiles and amphibians of Ohio. pp, 633–734, *in* Report of the Geological Survey of Ohio. Volume IV. Zoology and Botany. Part I. Zoology. Legislature of Ohio, Columbus, OH.

Smith, W.H. 1883. Bericht über die Reptilien und Amphibien von Ohio. pp, 691–801, 1109 (index) *in* Bericht über die Geologische Aufnahme von Ohio. IV. Band. Zoologie und Botanik. I. Theil. Zoologie. Gesetzgebung von Ohio, Columbus, OH

Walker, C.F. 1946. The Amphibians of Ohio: Part 1: The Frogs and Toads (Order Salientia). Ohio State Museum Science Bulletin 1(3): 1–109, frontispiece. [reprinted by Ohio Historical Society, Columbus, 1967].

Wilcox, E.V. 1891. Notes on Ohio batrachians. Otterbein Aegis 1(9):133–135. [reprinted 1961 by the Ohio Herpetological Society, 1961].

Wynn, D.E. and M. Moody. 2006. Ohio Turtle, Lizard, and Snake Atlas. Ohio Biological Survey, Columbus, OH. 80 pp.

Oklahoma

Altman, R.W. and G.R. Cline. 1985. Introduction to the Snakes of Oklahoma. Cooperative Extension Service, Division of Agriculture Oklahoma State University (9010). 6 pp.

Black, J.H. 1977. Endangered and threatened amphibians and reptiles of Oklahoma. Bulletin of Oklahoma Herpetological Society 2:42–50.

Black, J.H. 1980. Amphibians of Oklahoma: A checklist. Bulletin of Oklahoma Herpetological Society 4:78–80.

Black, J. and G. Sievert. 1989. A Field Guide to Amphibians of Oklahoma. Department of Wildlife Conservation, Oklahoma City, OK. 80 pp.

Bonn, E.W. and H.W. McCarley. 1953. The amphibians and reptiles of the Lake Texoma Area. Texas Journal of Science 5:465–471.

Bragg, A.N. 1941. A corrected list of the Amphibia known in central Oklahoma. Proceedings of the Oklahoma Academy of Sciences 22:16–17.

Bragg, A.N. 1952. Amphibians of McCurtain County, Oklahoma. Wasmann Journal of Biology 10:241–250.

Bragg, A.N. and William F. Hudson. 1951. New county records of Salientia and a summary of known distribution of Caudata in Oklahoma. Great Basin Naturalist 11:87–90.

Bragg, A.N., A.O. Weese, H.A. Dundee, H.T. Fisher, A. Richard, and C.B. Clark. 1950. Researches on The Amphibia of Oklahoma. University of Oklahoma Press, Norman, OK. 154 pp.

Bryce, F.D. 1975. Reptiles and amphibians of the Wichitas. Great Plains Journal 15:65–78.

Carpenter, C.C. and J. Krupa. 1989. Oklahoma Herpetology: An Annotated Bibliography. University of Oklahoma Press, Norman, OK. vii + 258 pp.

Cross, H. 1917. Animal and plant life of Oklahoma. Oklahoma Geological Survey 6. 68 pp. [amphibians and reptiles on pp. 21–23, 34–35].

Engbretson, G. 1974. Amphibians and Reptiles of Oklahoma. pp. 103–188 *in* P.G. Risser (ed.), Field Guide to Oklahoma. Oklahoma Biological Survey, Oklahoma City, OK.

Force, E.R. 1930. The amphibians and reptiles of Tulsa County, Oklahoma and vicinity. Copeia 1930:25–39.

Ginn, R. 1979. Poisonous snakes of Oklahoma. Bulletin of Oklahoma Herpetological Society 4:43–44.

Glass, B.P. 1975. Mammals, Reptiles and Amphibians. pp. 13–21 *in* Rare and Endangered Vertebrates and Plants of Oklahoma. US Department of Agriculture, Soil Conservation Service. 44 pp.

Lardie, R.L. 1977. Herpetology of Garfield County, Oklahoma. Special Publication of the Oklahoma Herpetological Society 1:1–26.

Lardie, R.L. 1982. A preliminary checklist of the amphibians and reptiles of northwestern Oklahoma (excluding the Oklahoma Panhandle). Bulletin of the Oklahoma Herpetological Society 7:36–78.

Lardie, R.L. and J.H. Black. 1981. The amphibians and reptiles of the Cimarron Gypsum Hills region in northeastern Oklahoma. Bulletin Oklahoma Herpetological Society 5: 76–125.

Murphy, J. 1987. Oklahoma lizards. Outdoor Oklahoma 43(3):30–35.

Ortenburger, A.I. 1925. Preliminary list of the snakes of Oklahoma. Proceedings of the Oklahoma Academy of Science 5:83–87.

Ortenburger, A.I. 1926. A report of the amphibians and reptiles of Oklahoma. Proceedings of the Oklahoma Academy of Science 6:89–100.

Ortenburger, A.I. 1927. A key to the snakes of Oklahoma. Proceedings of the Oklahoma Academy of Science 6:197–218.

Ortenburger, A.I. 1930. A key to the lizards and snakes of Oklahoma. Publication of the University of Oklahoma Biological Survey 2:209–239.

Overdeer, D. 1991. Reptiles of Mesquite Grasslands of Southwest Oklahoma. Bulletin Oklahoma Herp Society 15: 13-49.

Sievert, G. and L. Sievert. 1988. A Field Guide to Reptiles of Oklahoma. Department of Wildlife Conservation, Oklahoma City, OK. 96 pp.

Sievert, G. and L. Sievert. 2006. A Field Guide to Oklahoma's Amphibians and Reptiles, Second Edition Oklahoma Department of Wildlife Conservation, Oklahoma City, OK. 205 pp.

Sievert, G. and L. Sievert. 2011. A Field Guide to Oklahoma's Amphibians and Reptiles, Third Edition. Oklahoma Department of Wildlife Conservation, Oklahoma City, OK. 211 pp.

Sievert, G. and L. Sievert. 2021. **A Field Guide to Oklahoma's Amphibians and Reptiles, Fourth Edition**. Oklahoma Department of Wildlife Conservation, Oklahoma City, OK. 231 pp.

Van Vleet, A.H. 1902. Snakes of Oklahoma. Second Biennial Report of the Oklahoma Territory Department of Geology and Natural History (1901–1902): 167–173.

Webb, R.G. 1970. **Reptiles of Oklahoma**. University of Oklahoma Press, Norman, OK. xi + 370 pp.

Oregon

Corkran, C.C. and C. Thoms. 1996. Amphibians of Oregon, Washington, and British Columbia. Lone Pine Publishing, Edmonton, Alberta. 175 pp. [2nd edition 2006 (176 pp.), 3rd edition 2018 (176 pp.)].

Fitch, H.S. 1936. Amphibians and Reptiles of the Rogue River Basin. American Midland Naturalist 17(3)634-652.

Gordon, K. 1939. **The Amphibia and Reptilia of Oregon**. Oregon State Monographs: Studies in Zoology No. 1. 82 pp.

Leonard, W.P., H.A. Brown, L.L.C. Brown, K.R. McAllister, and R.M. Storm. 1993. Amphibians of Washington and Oregon. Seattle Audubon Society, Seattle, WA. vii + 168 pp.

Nussbaum, R.A., E.D. Brodie, Jr., and R.M. Storm. 1983. Amphibians and Reptiles of the Pacific Northwest. University Press of Idaho, Moscow, ID. 332 pp.

Pickwell, G. 1947. Amphibians and Reptiles of the Pacific States. Stanford University Press, Stanford, CA. xiv + 236 pp. [reprinted 1949; reprinted 1972 by Dover Publications, New York, NY, xviii + 234 pp. with a new foreword and table of changes in nomenclature by G. V. Pickwell].

Slevin, J.A. 1934. A Handbook of Reptiles and Amphibians of the Pacific States. California Academy of Sciences, San Francisco, CA. 73 pp.

St. John, A.D. 1980. Knowing Oregon Reptiles, Salem Audubon Society, Salem, OR. 36 pp.

St. John, A. 2002. **Reptiles of the Northwest**. Lone Pine Press, Renton, WA. 272 pp. [revised 2021].

Storm, R. 1974. Oregon snakes. Oregon Wildlife. August 1974:3–8.

Storm, RM. and WP. Leonard (eds). 1995. Reptiles of Washington and Oregon. Seattle Audubon Society, Seattle, WA. vii + 176 pp.

Pennsylvania

Allen, Jr., W.B. 1992. The snakes of Pennsylvania. Reptile and Amphibian Magazine, Pottsville, PA. 33 pp.

Baldauf, R.J. 1943. Handlist of Berks County amphibians and reptiles. Leaflets of the Mengel Natural History Society 1. 8 pp.

Barrett, J.A. 1944. Snakes in Pennsylvania. Pennsylvania Angler 13(1):10–12.

Conant, R. 1937. The snakes of the Philadelphia region. Frontiers 1(4):118–123.

Conant, R. 1940. Frogs and toads found near Philadelphia. Fauna 2(1):14–17.

Conant, R. 1942. Amphibians and Reptiles from Dutch Mountain (Pennsylvania) and Vicinity. American Midland Naturalist 27:154-170.

Conservation Education Division. 1979. Pennsylvania Reptiles and Amphibians. Pennsylvania Game Commission, Harrisburg, PA. 24 pp.

Fowler, H.W. 1915. An Annotated List of the Cold-Blooded Vertebrates of Delaware County, Pennsylvania. Proceedings of the Delaware County Institute of Science 7(2):33–45.

Gray, B.S. and M. Lethaby. 2008. The amphibians and reptiles of Erie County, Pennsylvania. Bulletin of the Maryland Herpetological Society. 44:49–69.

Harrison, H.H. 1949. Pennsylvania reptiles & amphibians. A picture story. Netting, M.G. (ed.).

(No. 1 .... Salamanders). Pennsylvania Angler 18(4):11.
(No. 2 .... Frogs). Pennsylvania Angler 18(5):10.
(No. 3 .... Snakes). Pennsylvania Angler 18(6):10.
(No. 4 .... Turtles). Pennsylvania Angler 18(7):10.
(No. 5 .... Snakes). Pennsylvania Angler 18(8):10.
(No. 6 .... Toads). Pennsylvania Angler 18(9):10.
(No. 7 .... Snapping Turtle. Pennsylvania Angler 18(10):10.
(No. 8 .... Salamanders). Pennsylvania Angler 18(11):24.
(No. 9 .... Blacksnakes). Pennsylvania Angler 18(12):14. 1950
(No. 10 ... Baby Turtles). Pennsylvania Angler 19(1):11.
(No. 11 ... Snakes). Pennsylvania Angler 19(2):24.
(No. 12 ... Frogs). Pennsylvania Angler 19(3):17.
(No. 13 ... Rattlesnakes). Pennsylvania Angler 19(4):18.
(No. 14 ... Mudpuppies and Hellbenders). Pennsylvania Angler 19(5):17.
(No. 15 ... Hunting Rattlesnakes). Pennsylvania Angler 19(6):16.
(No. 16 ... Salamanders). Pennsylvania Angler 19(7):16.
(No. 17 ... Water Snakes). Pennsylvania Angler 19(8):14.
(No. 18 ... Turtles). Pennsylvania Angler 19(9):12.
(No. 19 ... Lizards). Pennsylvania Angler 19(10):16.
(No. 20 ... Addenda). Pennsylvania Angler 19(11):16.

[reprinted with minor revisions in three editions and multiple printings under a single cover by the Pennsylvania Fish Commission, Harrisburg, PA; 1st edition (22 pp.), 2nd edition (24 pp. + text on inside covers), 3rd edition (24 pp. + text on inside covers); 2nd and 3rd editions with Netting, M.G. and Richmond, N.D. (eds.) and foreword by G.L. Trembley on inside cover; only 3rd edition printings bear a date, reprinted at least until 8th printing in 1976].

Hulse, A.C., C.J. McCoy, and E.J. Censky. 2001. **Amphibians and Reptiles of Pennsylvania and the Northeast**. Cornell Press, Ithaca, NY. xi + 419 pp.

McCoy, C.J. 1980. Identification Guide to Pennsylvania Snakes. Carnegie Museum of Natural History Education Bulletin No. 1. 12 pp.

McCoy, C.J. 1982. Amphibians and Reptiles in Pennsylvania. Carnegie Museum of Natural History Special Publication. No. 6. 91 pp.

McCoy, C.J. (ed.) 1985. Amphibians and Reptiles. pp. 257–295 *in* Genoways, H.H. and F.J. Brenner, eds. Species of Special Concern in Pennsylvania. Carnegie Museum of Natural History Special Publication No. 11. vi + 430 pp.

McCoy, C.J. 1986. Bibliography of Pennsylvania herpetology: 1981–1986. Proceedings of the Pennsylvania Academy of Science 60:122–124.

McCoy, C.J. 1992. Bibliography of Pennsylvania herpetology: 1987–1991. Proceedings of the Pennsylvania Academy of Science 66:94–96.

McKinstry, D.M. and S. Felege. 1974. Snakes of northwestern Pennsylvania. Bulletin of Maryland Herpetological Society 10:29–31.

Meshaka, W.E. and J.T. Collins. 2009. Pocket Guide to Pennsylvania Snakes. Pennsylvania Museum Commission, Harrisburg, PA. 53 pp.

Meshaka, W. E. and J.T. Collins. 2010. Pocket Guide to Pennsylvania Frogs and Toads. Pennsylvania Museum Commission, Harrisburg, PA. 40 pp.

Meshaka, W.E. and J.T. Collins. 2012. Pocket Guide to Lizards and Turtles of Pennsylvania. Pennsylvania Museum Commission, Harrisburg, PA. 40 pp.

Meshaka, W.E. and J.T. Collins. 2012. Pocket Guide to Salamanders of Pennsylvania. Pennsylvania Museum Commission, Harrisburg, PA. 51 pp.

Netting, M.G. 1930. The occurrence of lizards in Pennsylvania. Annals of the Carnegie Museum 19(3):169–174.

Netting, M.G. 1933. The amphibians of Pennsylvania. Proceedings of Pennsylvania Academy of Science 7:100–110.

Netting, M.G. 1933. The amphibians and reptiles of Pennsylvania. Pennsylvania Game News 4(1):12–15.

Netting, M.G. 1935. A non–technical key to the amphibians and reptiles of western Pennsylvania. Nawakwa Fireside, N.S. 3 Nos.3–4:34–49

Netting, M.G. 1936. Hand List of the Amphibians and Reptiles of Pennsylvania. Carnegie Museum Herpetology Leaflet, No. 1. 4pp.

Netting, M.G. 1939. Handlist of the Amphibians and Reptiles of Pennsylvania (second edition). Biennial Report of Pennsylvania Fish Commission for 1936–1938: 109–112.

Netting, M.G. 1939. The amphibians of Pennsylvania, (2nd edition). Biennial Report of Pennsylvania Fish Commission for 1936–1938: 113–122.

Netting, M.G. 1939. The reptiles of Pennsylvania. Biennial Report of Pennsylvania Fish Commission for 1936–1938:122–132.

Netting, M.G. 1946. The Amphibians and Reptiles of Pennsylvania. Pennsylvania Fish Commission, Harrisburg, PA. 29 pp. [reprinted 1949].

Palmer, T.C. 1908. A familiar talk on Delaware County Frogs. Proceedings of the Delaware

County Institute of Science 4(1):12–22.
Roddy, H.J. 1928. Reptiles of Lancaster County and the State of Pennsylvania. Science Press, Lancaster, PA. 53 pp.
Serrao, J. 2000. The Reptiles and Amphibians of the Poconos and Northeastern Pennsylvania. Llewellyn and McKane, Wilkes–Barre, PA. 48 pp.
Shaffer, L.L. 1991. **Pennsylvania Amphibians and Reptiles**. Pennsylvania Fish Commission, Harrisburg, PA. vii + 161 pp. [revised 1995, 1999].
Shiffer, C.N. n.d. Snakes in Pennsylvania. Pennsylvania Fish and Game, Harrisburg, PA. 4 pp. [reprinted many times].
Stewart, N.H. 1926. The Amphibia of Pennsylvania. Bucknell University, Lewisburg, PA. 16 pp.
Stewart, N.H. 1929. Reptiles of Pennsylvania. Bucknell University, Lewisburg, PA, 16 pp.
Surface, H.A. 1906. The serpents of Pennsylvania. Penn. Dept. of Agriculture Monthly Bulletin 4:114–208 + pls. 15–42.
Surface, H.A. 1907. The lizards of Pennsylvania. Penn. Dept. of Agriculture Monthly Bulletin 5:234–264 + pls. 30–33.
Surface, H.A. 1908. The turtles of Pennsylvania. Penn. Dept. of Agriculture Monthly Bulletin 6:106–196 + pls. 4–12.
Surface, H.A. 1913. The amphibians of Pennsylvania. Penn. Dept. of Agriculture Zoological Bulletin 3:67–151 + pls. 1–11.
Swanson, P.L. 1952. The reptiles of Venango County, Pennsylvania. American Midland Naturalist 47:161–182.
Wingard, R.G. 1958. Some Pennsylvania Snakes. The Pennsylvania State University, College of Agriculture, Extension Service Circular 482. 8 pp.

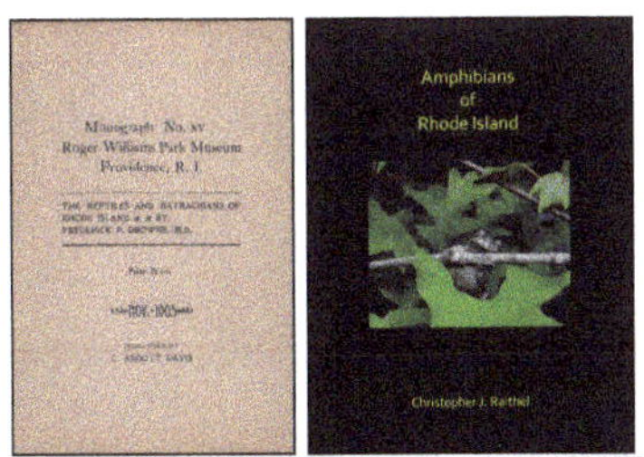

Rhode Island

Anonymous. 1918. The reptiles of Rhode Island. Part I – Turtles. Park Museum Bulletin 10:81–84.
Anonymous. 1918. The reptiles of Rhode Island. Part II – Snakes. Park Museum Bulletin 10:85–88. [Reprinted 1923].
Anonymous. 1918. The batrachians of Rhode Island, Part 1. Salamanders. Park Museum Bulletin. 10:89–92.
Anonymous. 1918. The batrachians of Rhode Island, Part 1. Toads and Frogs. Park Museum Bulletin. 10:93–96.
Anonymous. 2004. Native Snakes of Rhode Island. Rhode Island Department of Environmental Management, Providence, RI. 12 pp.
Anonymous. 2004. Turtles in Rhode Island. Rhode Island Department of Environmental Management, Providence, RI. 13 pp.
Bumpus, H. C. 1884 – 1886. Reptiles and Batrachians of Rhode Island.

Number I. Random Notes on Natural History 1(10):4–5. (Introduction)
Number II. Random Notes on Natural History 1(11):6–7. (Order Testudinata, Part 1)
Number III. Random Notes on Natural History 1(12):6–7. (Order Testudinata, Part 2)
Number IV. Random Notes on Natural History 2(1):5. (Order Testudinata, Part 3)
Number V. Random Notes on Natural History 2(2):13. (Order Sauria – lizards)
Number VI. Random Notes on Natural History 2(3):20–21. (Order Sauria – snakes introduction)
Number VII. Random Notes on Natural History 2(4):28. (Order Sauria – snakes, Part 2)
Number VIII. Random Notes on Natural History 2(5):37–38. (Order Sauria – snakes, Part 3)
Number IX. Random Notes on Natural History 2(6):44–45. (Order Sauria – snakes, Part 4)
Number X. Random Notes on Natural History 2(7):52–53. (Order Anura, Part 1)
Number XI. Random Notes on Natural History 2(8):59–60. (Order Anura, Part 2)
Number XII. Random Notes on Natural History 2(9):68–69. (Order Anura, Part 3)
Number XIII. Random Notes on Natural History 2(10):75. (Order Anura, Part 4)
Number XIV. Random Notes on Natural History 2(11):83–84. (Order Anura, Part 5)
Number XV. Random Notes on Natural History 2(12):93. (Order Anura, Part 6)
Number XVI. Random Notes on Natural History 3(1):7–8. (Order Anura, Part 7)
Number XVII. Random Notes on Natural History 3(2):13. (Order Anura, Part 8)
Number XVIII. Random Notes on Natural History 3(3):21. (Order Urodela, Part 1)
Number XIX. Random Notes on Natural History 3(5):35. (Order Urodela, Part 2)
Number XX. Random Notes on Natural History 3(6):43. (Order Urodela, Part 3)
Number XXI. Random Notes on Natural History 3(7):52. (Order Urodela, Part 4)
Number XXII. Random Notes on Natural History 3(9):69. (Order Urodela, Part 5)
Number XXIII. Random Notes on Natural History 3(10):76. (Order Urodela, Part 6)
Number XXIV. Random Notes on Natural History 3(11):83–84. (Order Urodela, Part 7)

Drowne, F.P. 1905. **The Reptiles and Batrachians of Rhode Island**. Roger Williams Park

Museum Monograph 15. 24 pp.

Raithel, C.J. 2019. **Amphibians of Rhode Island**. Rhode Island Division of Fish and Wildlife, West Kingston, RI. 316 pp.

Woodruff, R.E. 1960. Checklist of Rhode Island reptiles & amphibians. The Narragansett Naturalist 3(2):55–56.

South Carolina

Beane, J., A.L. Braswell, J.C. Mitchell, W.M. Palmer, and J. R. Harrison. 2010. Amphibians and Reptiles of the Carolinas and Virginia. University of North Carolina Press, Chapel Hill, NC. 274 pp.

Camper, J.D. 2019. **The Reptiles of South Carolina**. University of South Carolina Press, Columbia, SC. 270 pp.

Chamberlain, E.B. 1933. The Charleston Museum List of South Carolina Reptiles. The Charleston Museum Leaflet 4. 7 pp.

Chamberlain, E.B. 1933. Frogs and Toads of South Carolina. The Charleston Museum Leaflet 12. 38 pp.

Corrington, J.D. 1929. Herpetology of the Columbia, South Carolina region. Copeia 172:58–83.

Freeman, H.W. 1955. Amphibians and reptiles of the Savannah River Plant area, Caudate Amphibia. University of South Carolina Publication Series III. 1:227–238.

Freeman, H.W. 1955. Amphibians and reptiles of the Savannah River Plant area, Chelonia. University of South Carolina Publication Series III. 1:239–244.

Freeman, H.W. 1955. Amphibians and reptiles of the Savannah River Plant area, Crocodilia, Sauria, and Serpentes. University of South Carolina Publication Series III. 1:275–291.

Freeman, H.W. 1956. Amphibians and reptiles of the Savannah River Plant area, Salientia. University of South Carolina Publication Series III. 2:26–35.

Gee, G. 1936. South Carolina vertebrate fauna [salamanders]. The Index–Journal, Greenwood, S.C., 4 October 1936:3.

Gee, G. 1936. South Carolina vertebrate fauna [frogs and toads]. The Index–Journal, Greenwood, S.C., 10 October 1936:9.

Gibbons, J.W. 1977. Snakes of the Savannah River Plant with Information about Snakebite Prevention and Treatment. SRO–NERP–1. 26 pp.

Gibbons, J.W. and J.R. Harrison III. 1981. Reptiles and amphibians of Kiawah and Capers Islands, South Carolina. Brimleyana 5:145–162.

Gibbons, J.W. and K.K. Patterson. 1978. The Reptiles and Amphibians of the Savannah River Plant SRO–NERP–2. 24 pp.

Gibbons, J.W. And R.D. Semlitsch. 1991. **Guide to the Reptiles and Amphibians of the Savannah River Site**. University of Georgia Press, Athens, GA. xii + 131 pp.

[paperback edition 2009].

Gibbons, J.W. and P.J. West (eds.) 1998. Snakes of Georgia and South Carolina. Savannah River Ecology Laboratory Herp Outreach Publication 1. 28 pp.

Gibbons, J.W., D.H. Nelson, K.K. Patterson, and J.L. Greene. 1976. The Amphibians and reptiles of the Savannah River Plant in west-central South Carolina. pp. 133–143 *in* Proceedings of the First South Carolina Endangered Species Symposium. Charleston, SC. 201 pp.

Hall, R.J. 1994. Herpetofaunal Diversity of the Four Holes Swamp, South Carolina. USDI National Biological Survey Resource Pub 198, Washington DC. iv + 43 pp.

Mancke, R. 1977. Common Snakes of South Carolina. South Carolina Museum Commission, Columbia, SC. 24 pp.

Martof, B.S., W.M. Palmer, J.R. Bailey, and J.R. Harrison. 1980. Amphibians and Reptiles of the Carolinas and Virginia. University of North Carolina Press, Chapel Hill, NC. [viii] + 264 pp. [undated softback reprint].

Montanucci, R.R. 2006. A review of the amphibians of the Jim Timmerman Natural Resources Area, South Carolina. Monograph 1, Southeastern Naturalist 5:1– 58.

Penney, J.T. 1952. Distribution and bibliography of the amphibians and reptiles of South Carolina. University of South Carolina Publications in Biology. 1:1–28.

Sanders, A.E. 1966. The reptiles of Columbia, S.C. and vicinity. Columbia Science Museum Quarterly, Summer 1966. iv + 36 pp.

Thompson, Jr., E.F. 1982. A Guide to the Amphibians, Reptiles, and Mammals of South Carolina. Privately Published, Columbia, SC. 134 pp

South Dakota

Ballinger, R.E., J.W. Meeker, and M. Thies. 2000. A checklist and distribution maps of the amphibians and reptiles of South Dakota. Transactions of Nebraska Academy of Science 26: 29–46.

Bandas, S.J. and K.F. Higgins. 2004. A Field Guide to South Dakota Turtles. South Dakota State University, Brookings, SD. 36 pp.

Chace, G.E. 1971. The Reptiles of Paha Sapa. A herpetologist's report from the Black Hills of South Dakota. Animal Kingdom 74:20–28.

Coues, E. and H.C. Yarrow. 1878. Notes on the herpetology of Dakota and Montana. Bulletin of the United States Geological Survey 4:259–291. [reprinted 1978 *in* Adler, K. (ed.), Herpetological Explorations of the Great American West, Volume 1. Arno Press, New York, NY].

Fischer, T.D., D.C. Backland, K.F. Higgins, and D.E. Naugle. 1999. Field Guide to South Dakota Amphibians. South Dakota Agricultural Experiment Station Bulletin 733. South Dakota State University, Brookings, SD. 52 pp.

Fishbeck, D.W. and J.C. Underhill. 1959. A checklist of the amphibians and reptiles of South Dakota. Proceedings of the South Dakota Academy of Science. 38:107–113.

Fishbeck, D.W. and J.C. Underhill. 1960. Amphibians of eastern South Dakota. Herpetologica 16:131–136.

Kiesow, A.M. 2006. Field Guide to Amphibians and Reptiles of South Dakota. South Dakota Department of Game, Fish and Parks, Pierre, SD. 178 pp. + CD.

Kiesow, A.M. and D.R. Davis. 2020. **Field Guide to Amphibians and Reptiles of South Dakota**. South Dakota Department of Game, Fish and Parks, Pierre, SD. 161 pp.

Over, W.H. 1923. **Amphibians and reptiles of South Dakota**. University of South Dakota Geological and Natural History Survey Bulletin 12. 34 pp. + 18 pls.

Over, W.H. 1943. Amphibians and Reptiles of South Dakota. South Dakota Natural History Studies VI, Univ. of South Dakota, Vermillion. 31 pp.

Reagan, A.B. 1908. Animals, reptiles and amphibians of Rosebud Indian Reservation, South Dakota. Transactions of the Kansas Academy of Sciences 21:163–165. [first published in German in 1907 as Säugetiere, Reptilien und Amphibien vom Rosebud-Indianer-Reservatgebiet in Süd-Dakota. Zoologischer Anzeiger 32:31–32].

Thompson, S.W. 1982. Snakes of South Dakota. South Dakota Conservation Digest 49(4):12–18.

Thompson, S.W. Turtles of South Dakota. South Dakota Conservation Digest 49(3):12–15.

Thompson, S. and D. Backlund. 1999. South Dakota Snakes: A Guide to Snake Identification. South Dakota Department of Game, Fish and Parks. Pierre, SD. 28 pp.

Tennessee

Anonymous. 1957. Awkward knights in armor: The turtles of Tennessee. The Tennessee Conservationist 23(5):18–23.

Blanchard, F.N. 1922. The amphibians and reptiles of western Tennessee. Occasional Papers of the Museum of Zoology at University of Michigan 177. 18 pp.

Dodd, Jr., K.C. 2004. The Amphibians of Great Smoky Mountains National Park. University of Tennessee Press, Knoxville, TN. 283 pp.

Eagar, D.C. and R.M. Hatcher (eds). 1980. Tennessee's Rare Wildlife, Volume I: The Vertebrates. Tennessee Wildlife Resources Agency, Nashville, TN. viii +338 pp. [amphibians and reptiles on pp. D1–D33].

Gentry, G. 1955–1956. An annotated check list of the amphibians and reptiles of Tennessee. Journal of Tennessee Academy of Science 30[1955]:168–176,

31[1956]:242–251.

Gentry, G. 1955. Frogs and toads. Tennessee Conservationist 21(7):16–17.

Gentry, G, R. Sinclair, W. Hon, and B. Ferguson. 1965. **Amphibians and Reptiles of Tennessee.** Tennessee Game and Fish Commission, Nashville, TN. 28 + [1] pp. [revised 1970; a compilation of multiple individual articles by the authors from The Tennessee Conservationist, including Anonymous (1957), see above].

Holtzclaw, F.W. and J.G. Byrd. 1979. Snakes of Tennessee. Tennessee Conservationist 45(3):8–11, (4):6–9.

Huheey, J.E. and A. Stupka. 1967. Amphibians and Reptiles of Great Smoky Mountains National Park. University of Tennessee Press, Knoxville, TN. ix + 98 pp.

King, W. 1939. A survey of the herpetology of Great Smoky Mountains National Park. American Midland Naturalist 21:531–582.

Niemiller, M.L., and R. G. Reynolds (eds). 2011. **The Amphibians of Tennessee.** University of Tennessee Press, Knoxville, TN. xxii + 369 pp.

Niemiller, M.L., R.G. Reynolds, and B.T. Miller (eds). 2013. The Reptiles of Tennessee. University of Tennessee Press, Knoxville, TN. xxiv + 366 pp.

Niemiller, M.L., R.G. Reynolds, B.M. Glorioso, J. Spiess, and B.T. Miller. 2011. Herpetofauna of the Cedar Glades and associated habitats of the Inner Central Basin of Middle Tennessee. Herpetological Conservation and Biology 6(1):127–141.

Parker, M.V. 1939. The amphibians and reptiles of Reelfoot Lake and vicinity, with a key for the separation of species and subspecies. Journal of the Tennessee Academy of Science 14:72–101.

Parker, M.V. 1948. A contribution to the herpetology of western Tennessee. Journal of the Academy of Science. 22 (1):20–30.

Redmond, W.H., A.C. Echternacht, and A.F. Scott. 1990. Annotated Checklist of Bibliography of Amphibians and Reptiles of Tennessee (1835 through 1989). Austin Peay State University Miscellaneous Publications of the Center Field Biology No.4. iii + 173 pp.

Redmond, W.H. and A.F. Scott. 1996. Atlas of Amphibians in Tennessee. Austin Peay State University Miscellaneous Publications of the Center Field Biology No.12. v + 94 pp.

Rhoads, S.N. 1895. Contributions to the zoology of Tennessee, No. 1: Reptiles and amphibians. Proceedings of the Academy of Natural Sciences of Philadelphia 47:376–407.

Scott, A.F. and W.H. Redmond. 2002. Updated Checklist of Tennessee's Amphibians and Reptiles with an Annotated Bibliography covering Primarily years 1990 through 2001. Center for Field Biology, Austin Peay State University, Clarksville, TN. 64 pp.

Scott, A.F., and W. H. Redmond. 2016. Atlas of Reptiles in Tennessee. The Center of Excellence for Field Biology, Austin Peay State University, Miscellaneous Publication Number 18. vi + 188 pp.

Snyder, D.H. 1972. Amphibians and Reptiles of Land Between the Lakes. Tennessee Valley Authority, Golden Pond, KY. 90 pp.

Snyder, D.H., A.F. Scott, E. Zimmerer, and D. Frymire. 2016. Amphibians and Reptiles of Land Between the Lakes. University Press of Kentucky, Lexington, KY. 107 pp.

Tilley, S.G. and J.E. Huheey. 2001. Reptiles and Amphibians of the Smokies. Great Smoky Mountains Natural History Association, Gatlinburg, TN. 143 pp.

Weller, W.H. 1931. A preliminary list of the salamanders of the Great Smoky Mountains of North Carolina and Tennessee. Proceedings of the Junior Society of Natural Sciences 2:21–32. [reprinted 1965 by the Ohio Herpetological Society, pp. 25–36 *in* Herpetological Papers from the Proceedings of the Junior Society of Natural Sciences (1930–1932)].

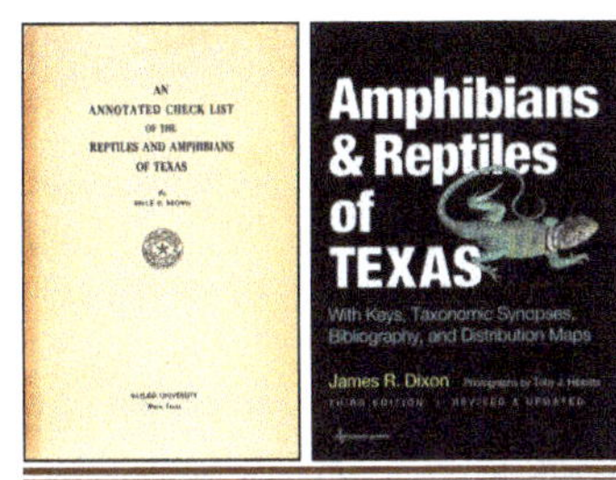

Texas

Axtell, R.W. 1986–2005. Interpretive Atlas of Texas Lizards. Privately published. [30 separately numbered and paginated accounts totaling ii + 559 pp., with 16 maps in loose-leaf, 3-hole-punched format].

Bartlett, R.D. and P. Bartlett. 1999. A Field Guide to Texas Reptiles and Amphibians. Gulf Publishing Company, Houston, TX. 330 pp.

Bassett, L. 2023. Updated Geographic Distributions for Texas Amphibians. Amphibians and Reptiles 30: e18486. 21 pp.

Bonn, E.W. and H.W. McCarley. 1953. The amphibians and reptiles of the Lake Texoma area. Texas Journal of Science 5:465–471.

Brown, B.C. 1950. **An Annotated Checklist of the Reptiles and Amphibians of Texas**. Baylor University Press, Waco, TX, xii + 257 pp.

Burt, C.E. 1938. Contributions to Texas herpetology VII. The Salamanders. American Midland Naturalist 20:374–380.

Chaney, A.H. 1982. Keys to the Vertebrates of Texas. Texas A&M University, Kingsville, TX. v + 99 pp. [revised editions in 1990 ("(Exclusive of Birds)" added to title, iii + 99 pp.), 1993 (iii + 101 pp.), 1996 ([1] + iii + 102 pp.)].

Crouse, H.W. 1902. Venomous Snakes and Spiders of Texas. Pp. 144–175, 9 pls. *in* Transactions of the State Medical Association of Texas. Thirty-fourth Annual Session, Held at Dallas, Texas, May 6, 7, 8 and 9, 1902. Von Boeckmann, Schutze & Co. Austin, TX. [also distributed as 39 pp. unpaginated reprint].

Davenport, J.W. 1943. Field Book of the Snakes of Bexar County, Texas and Vicinity. Publication of Witte Memorial Museum, San Antonio, TX. 132 pp.

Davis, D.R. and T.J. LaDuc. 2018. Amphibians and reptiles of C.E. Miller Ranch and the Sierra Vieja, Chihuahuan Desert, Texas, USA. ZooKeys 735:97-130.

Dayton, G.H., R. Skiles, and L. Dayton. 2007. Frogs and Toads of Big Bend National Park. Texas A&M University Press, College Station, TX. 51 pp.

Dixon, J.R. 1987. Amphibians and Reptiles of Texas. Texas A&M Press, College Station, TX. xii+ 424 pp.

Dixon, J.R. 1993. Supplement to the Literature for the "Amphibians and Reptiles of Texas" 1987. Smithsonian Herpetological Information Service 94. 43 pp.

Dixon, J.R. 1996. Ten Year Supplement to Texas Herpetological County Records Published in "Amphibians and Reptiles of Texas, 1987". Texas Herpetological

Society Special Publication 2. 64 pp.

Dixon, J.R. 2000. Amphibians and Reptiles of Texas, second edition. Texas A&M University Press, College Station, TX. 421 pp.

Dixon, J.R. 2013. **Amphibians and Reptiles of Texas, third edition**. Texas A&M University Press, College Station, TX. viii + 447 pp.

Dixon, J.R. 2015. Herpetofauna of Texas / Herpetofauna de Texas. Pp. 181–205, 399–413 *in* Lemos-Espinal, J.A. (ed.), Amphibians and Reptiles of the US-Mexico Border States / Anfibios y Reptiles de los Estados de la Fontera México–Estados Unidos. Texas A&M University Press, College Station, TX. [checklist, plates and index collective for entire book].

Dixon, J.R. and Smith, C.R. (eds.). 1999. Amphibians and Reptiles of Texas. Taxonomic and Distributional Inventory with Bibliography. Texas System of Natural Laboratories Index Series no. HTX–99: xviii + 519 pp. Texas System of Natural Laboratories, Inc. Austin TX.

Dixon, J.R. and J.E. Werler. 2005. Texas Snakes – A Field Guide. University of Texas Press, Austin, TX. 364 pp.

Dixon, J.R., J.E. Werler, and M.R.J. Forstner. 2020. Texas Snakes. A Field Guide. Revised Edition. University of Texas Press, Austin, TX. xvi + 448 pp.

Garrett, J.M. and D.G. Barker. 1987. A Field Guide to Reptiles and Amphibians of Texas. Texas Monthly Press, Austin, TX. xi + 225 pp.

Gloyd, H.K. 1944. Texas snakes. Texas Geographic Magazine 8:1–18.

Hardy, L.M. 1979. Checklist of the Amphibians and Reptiles of Caddo Lake Watershed of Texas and Louisiana. Bulletin of Museum of Life Science, Louisiana State University 10. 31 pp.

Hibbitts, T.D. and T.L. Hibbitts. 2015. Texas Lizards: A Field Guide. University of Texas Press, Austin, TX. 333 pp.

Hibbitts, T.D. and T.L. Hibbitts. 2016. Texas Turtles and Crocodilians: A Field Guide. University of Texas Press, Austin, TX. 257 pp.

Huang, T.T., Lewis, S.R. and Lucas, B.S, III 1975. Venomous snakes 121–164 *in* M.D. Ellis, (ed) Dangerous Plants, Snakes, Arthropods & Marine Life of Texas. US Department of Health, Education, and Welfare, Public Health Service. Galveston, TX.

Jameson, D.L. and A.G. Flury. 1949. The reptiles and amphibians of the Sierra Vieja Range of southwestern Texas. Texas Journal of Science 1:54–79.

Mitchell, J.D. 1903. The poisonous snakes of Texas, with notes on their habits. Transactions of the Texas Academy of Science 5:19–48 [also published in The Texas Medical News 12:411–437].

Mosauer, W. 1932. The Amphibians and Reptiles of the Guadalupe Mountains of New Mexico and Texas. Occasional Papers of the Museum of Zoology University of Michigan. 246. 18 pp. + 1 pl.

Murrary, L.T. 1939. Annotated list of amphibians and reptiles from the Chisos Mountains. Contribution from Baylor University Museum 24:4–16.

Price, A.H. 1998. Poisonous Snakes of Texas. Texas Parks and Wildlife Press. Austin, TX. 112 pp.

Price, A.H. 2009. Venomous Snakes of Texas. University of Texas Press, Austin, TX. 116 pp.

Price, M. 2022. Reptiles of the Trans-Pecos Texas. ECO Publishing, Rodeo, NM. 210 pp.

Prival, D. and M. Goode. 2011. Chihuahuan Desert National Parks Reptile and Amphibian Inventory. National Park Service. Natural Resources Technical Report NPS/CHDN/NRTR–2011/489, Fort Collins, CO. 90 pp.

Raun, G.G. 1965. A Guide to Texas Snakes. Texas Memorial Museum, Museum Notes No. 9. 85 pp.

Raun, G.G. and F.R. Gehlbach. 1972. Amphibians and Reptiles in Texas. Dallas Museum of Natural History Bulletin 2. 61 pp.

Strecker, J.K. 1908. The reptiles and batrachians of Victoria and Refugio counties, Texas. Proceedings of the Biological Society of Washington 21:47–52.

Strecker, J.K. 1908. A preliminary annotated list of Batrachia of Texas. Proceedings of the Biological Society of Washington 21:53–61.

Strecker, J.K. 1908. The reptiles and amphibians of McLennan County, Texas. Proceedings of the Biological Society of Washington 21:69–83.

Strecker, J.K. 1909. Reptiles and amphibians collected in Brewster County, Texas. Baylor University Bulletin 12:11–15.

Strecker, J.K. 1915. Reptiles and Amphibians of Texas. Baylor University Bulletin 18. 82 pp.

Strecker, J.K. 1922. An annotated catalogue of the amphibians and reptiles of Bexar County, Texas. Scientific Society of San Antonio Bulletin 4:1–31.

Strecker, J.K. 1926. A list of the reptiles and amphibians collected by Louis Garni in the vicinity of Boerne, Texas. Contributions from Baylor University Museum. 6:3–9.

Strecker, J.K. 1926. Amphibians and reptiles collected in Somerwell County, Texas. Contributions from Baylor University Museum. 2:1–2.

Strecker, J.K. 1926. Chapters from the life-histories of Texas reptiles and amphibians. 1. Contributions from Baylor University Museum. 8:2–12.

Strecker, J.K. 1926. Notes on the herpetology of the East Texas Timber Belt. 1. Liberty County amphibians and reptiles. Contributions from Baylor University Museum. 3:1–3.

Strecker, J.K. 1926. Notes on the herpetology of the East Texas Timber Belt. 2. Henderson County amphibians and reptiles. Contributions from Baylor University Museum. 7:3–7.

Strecker, J.K. 1926. Reptiles from Lindale, Smith County, Texas. Contributions from Baylor University Museum 7:7.

Strecker, J.K. 1927. Chapters from the life-histories of Texas reptiles and amphibians. 2. Contributions from Baylor University Museum 10:1–14.

Strecker, J.K. 1928. Amphibians and reptiles collected at Harlingen, Texas. Contributions from Baylor University Museum. 15:7–8.

Strecker, J.K. 1929. A preliminary List of the amphibians and reptiles of Tarrant County, Texas. Contr. Baylor Univ. Mus. Waco 19:10–15.

Strecker, J.K. 1929. Field notes on the herpetology of Wilbarger County, Texas. Contr. Baylor Univ. Mus. Waco 19:1–9.

Strecker, J.K. 1930. A catalog of the amphibians and reptiles of Travis County, Texas. Contributions from Baylor University Museum 23:1–16.

Strecker, J.K.and J.E. Johnson. 1935. Notes on the herpetology of Wilson County, Texas. Baylor University Bulletin 38(3):17–23.

Strecker, J.K. and W.J. Williams. 1927. Herpetological records from the vicinity of San Marcos, Texas, with distributional data on the amphibians and reptiles of the Edwards Plateau region and central Texas. Contributions to Baylor University

Museum 12:1–16.

Strecker, J.K. and W.J. Williams. 1928. Field notes on the herpetology of Bowie County, Texas Contributions from Baylor University Museum 17:1–16.

Tennant, A. 1984. The Snakes of Texas. Texas Monthly Press, Austin, TX. 561 pp.

Tennant, A. 1985. A Field Guide to Texas Snakes. Texas Monthly Press, Austin, TX. 260 pp.

Tennant, A. 1998. A Field Guide to Texas Snakes, Second Edition. Gulf Publishing, Houston TX. xix + 291pp. plus folded poster in pocket.

Tennant, A. 2005. Field Guide to Texas Snakes, Third Edition. Lone Star Guides, Taylor Publishing, Lanham, MD. 352 pp.

Thomas, R.A. 1974. A Checklist of Texas Amphibians and Reptiles. Texas Parks and Wildlife Department Technical Series No. 17. 15 pp. [revised 1976, iv + 16 pp.].

Tipton, B.L., T. Hibbitts, T. Hibbitts, T. Hibbitts, and T. LaDuc. 2012. Texas Amphibians: A Field Guide. University of Texas Press, Austin, TX. xiv + 309 pp.

Vermersch, T. 1992. Lizards and Turtles of South–Central Texas. Eakin Press, Austin, TX. xiv + 170 pp.

Vermersch, T.G. and R.E. Kunz. 1986. Snakes of South–Central Texas. Eakin Press, Austin, TX. xiii + 137 pp.

Werler, J.E. 1950. Poisonous snakes of Texas and the first aid treatment of their bites. Texas Game and Fish 8(3):25–41. [reprinted 1950 (16 pp.), reprinted as Texas Parks and Wildlife Department Bulletin 31 in 1950 (40 pp.), revised 1952 (40 pp.), reprinted 1955, 1958, 1960, revised 1963 (62 pp.), revised 1964 (62 pp.), reprinted 1967, 1970].

Werler, J.E. 1978. Poisonous snakes of Texas. Texas Parks and Wildlife Department Bulletin 31. iv + 53 pp.

Werler, J.E. and J.R. Dixon. 2000. Texas Snakes. University of Texas Press, Austin, TX. 437 pp.

Wright, A.H. and A.A. Wright. 1938. Amphibians of Texas. Proceedings of the Transactions of the Texas Academy of Science 21:5–44.

Utah

Anonymous. 1987. Utah's Amphibians and Reptiles. Utah Division of Wildlife Resources., Salt Lake City, UT. 12 pp.

Baird, S.F. and C. Girard. 1852. Appendix C.— Reptiles. pp. 336–353, pls. 1–6 *in* Stansbury, H. Exploration and Survey of the Valley of the Great Salt Lake of Utah, Including a Reconnoissance of a New Route Through the Rocky Mountains. Senate of the United States Executive [Document] No. 3. Lippincott, Grambo & Co., Philadelphia, PA. [several early editions, see Banta, B. 1967. The editions of the Stansbury Report. The Pan-Pacific Entomologist 43:307–308; reprinted

1978 *in* Adler, K. (ed.), Herpetological Explorations of the Great American West, Volume 1. Arno Press, New York, NY and 1988 by Smithsonian Institution Press, Washington, DC].

Bartholomew, B. 1993. Key to the amphibians of Utah. Intermontanus 2(3):3–4.

Bartholomew, B. 1993. Checklist of the amphibians & reptiles of Utah. Utah Association of Herpetologists. Unpaginated. [5 pp.].

Cowles, R.B. and C.M. Bogert. 1936. The herpetology of the Boulder Dam region. Herpetologica 1:33–42.

Cox, D.C. and W.W. Tanner. 1995. **Snakes of Utah**. Brigham Young University. Provo, UT. 92 pp.

Lentsch, L. D., M. J. Perkins, P. D. Thompson, J. J. Wallace, and S. C. Hansen. 1995. Native Fish, Amphibians, & Reptiles of Utah. State of Utah Natural Resources, Division of Wildlife Resources, 95-20. iv + 147 pp.

Pack, H.J. 1930. Snakes of Utah. Bulletin of the Utah Agricultural Experiment Station 221. 32 pp.

Rawly, E.V. 1970. Snake Distribution of Utah. Utah Division of Wildlife Resources, Salt Lake City, UT. 22 pp.

Schwinn, M.A. and L. Minden. 1979. Utah Reptile and Amphibian Latilong Distribution. Utah Division of Wildlife Resources, Salt Lake City, UT. 80–1. viii + 10 pp.

Shofner, R.M. 2007. A Modern Checklist of the Amphibians, Reptiles and Turtles of Utah. Kansas Herpetological Society, Lawrence, KS. 3 pp.

Tanner, V.M. 1927. Distributional list of the amphibians and reptiles of Utah, No. 1. Copeia 163:54–58.

Tanner, V.M. 1928. Distributional list of the amphibians and reptiles of Utah, No. 2. Copeia 166:22– 28

Tanner, V.M. 1929. Distributional list of the amphibians and reptiles of Utah, No. 3. Copeia 172:46–52

Tanner, V.M. 1931. A synoptical study of Utah Amphibia. Utah Academy of Sciences 8:159–198 + pls. 8–20.

Tanner, W.W. 1939. Reptiles of Utah County. Proceedings of the Utah Academy of Science, Arts and Letters 16:105. [lists 22 species, names only].

Tanner, W.W. 1975. 1975 Checklist of Utah amphibians and reptiles. Proceedings of the Utah Academy of Sciences, Arts, and Letters. 52:4–8.

Van Denburgh, J. and J.R. Slevin. 1915. A list of the amphibians and reptiles of Utah, with notes on the species in the collection of the Academy. Proceedings of the California Academy of Sciences, 4th Series 5:99–110.

Wauer, R.H. 1964. Reptiles and Amphibians of Zion National Park. Zion Natural History Association, Zion National Parks, UT. 55 pp.

Woodbury, A.M. 1928. The reptiles of Zion National Park. Copeia (166):14–21.

Woodbury, A.M. 1931. **The Reptiles of Utah**. Bulletin of the University of Utah 21(5):1–129.

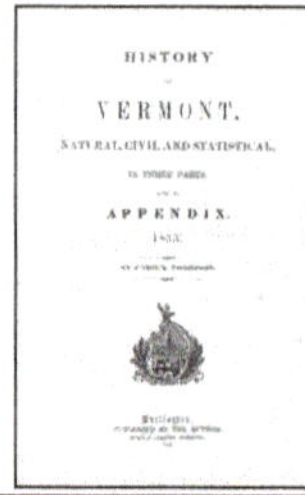

Vermont

Andrews, J.S. 1995. A Preliminary Atlas of the Reptiles and Amphibians of Vermont. Vermont Reptiles and Amphibian Scientific Advisory Group. Middlebury, VT. 64 pp.

Andrews, J.S. 2002. **The Atlas of the Reptiles and Amphibians of Vermont**. Vermont Department of Fish and Game, Waterbury, VT. iv + 90 pp. [2013 Update (80 pp.), 2019 Update (74 pp.), updates published by The Vermont Reptile and Amphibian Atlas, Salisbury, VT].

Anonymous. n.d. Amphibians of Vermont. Vermont Department of Fish and Game, Montpelier, VT 3 pp.

Thompson, Z. 1842. **History of Vermont, Natural, Civil, and Statistical, in Three Parts, with a New Map of the State and 200 Engravings**. Published for the Author by Chauncey Goodrich, Burlington, VT. iv + 224 + 224 + 200 + [iv] pp., 1 folding map. [herpetological section is Chapter IV. Reptiles of Vermont pp. 112–127 in Part First. Natural History of Vermont; revised in 1853 as History of Vermont, Natural, Civil, and Statistical, in Three Parts, with an Appendix. Published by the Author, Burlington, VT. iv + 224 + 224 + 200 + 64 + iv pp., 1 folding map (herpetological section as above + pp.28–30 in Appendix; Natural History of Vermont only reprinted with new front matter in 1972 (reprinted 1982) by Charles E. Tuttle, Rutland, VT. xii + 283 + [1] + [iv] pp. (in this facsimile, 1853 Appendix renumbered 225–283 to follow continuously from original 1842 pagination of Part First)].

Williams, S. 1794. The Natural and Civil History of Vermont. Isaiah Thomas and David Carlisle, Walpole, New Hampshire. xvi + 416 pp, 1 folding map. [amphibians and reptiles on pp. 125–129].

Virginia

Beane, J., A.L. Braswell, J.C. Mitchell, W.M. Palmer, and J.R. Harrison. 2010. Amphibians and Reptiles of the Carolinas and Virginia. University of North Carolina Press, Chapel Hill, NC. 274 pp.

Bierly, E.J. 1954. Turtles in Maryland and Virginia. Atlantic Naturalist 9:244–249.

Brothers, D.R. 1992. An Introduction to the Snakes of the Dismal Swamp Region of North Carolina and Virginia. Edgewood PROBES, Inc., Boise, ID. viii + 139 pp. + 4 pls.

Burch, P.R. 1940. Snakes of the Alleghany Plateau of Virginia. Virginia Journal of Science 1:35–40.

Burger, W. L. 1958. List of Virginian amphibians and reptiles. Virginia Herpetological Society Bulletin 4 (supplement):1–5.

Conant, R. 1945. An Annotated Checklist of the Amphibians and Reptiles of the Del-Mar-Va Peninsula. Society of Natural History of Delaware, Wilmington, DE. 8 pp.

Conant, R., J.C. Mitchell, and C.A. Pague. 1990. Herpetofauna of the Virginia Barrier Islands. Virginia Journal of Science 41:364–380.

Ernst, C.H., S.C. Belfit, S.W. Sekscienski, and A.F. Laemmerzahl. 1997. The amphibians and reptiles of Ft. Belvoir and Northern Virginia. Bulletin of the Maryland Herpetological Society 33:1–62.

Dunn, E.R. 1918. A preliminary list of the reptiles and amphibians of Virginia. Copeia (53):16–27.

Hardy, Jr., J.D. 1972. Amphibians of the Chesapeake Bay region. Chesapeake Science 13(supplement):123–128.

Hardy, Jr., J.D. 1972. Reptiles of the Chesapeake Bay region. Chesapeake Science 13 (supplement):128–134.

Hay, W.P. 1902. A list of the batrachians and reptiles of the District of Columbia and vicinity. Proceedings of the Biological Society of Washington 15:121–145.

Hoffman, R.L. 1949. The turtles of Virginia. Virginia Wildlife 10(August):16–19.

Kleopfer, J.D., T.S.B. Akre, S.H. Watson, and R. Boettcher. 2014. A Guide to the Turtles of Virginia. Virginia Department of Game and Inland Fisheries, Richmond, VA. 44 pp. [revised 2022 (48 pp.)].

Kleopfer, D. and C.S. Hobson. 2011. A Guide to the Frogs and Toads of Virginia. Virginia Department of Game and Inland Fisheries, Richmond, VA. 44 pp. + CD.

Kleopfer, J.D., C.S. Hobson S. H. Watson, and S.M. Roble. 2021. A Guide to the Frogs and Toads of Virginia, Second Edition. Virginia Department of Game and Inland Fisheries, Richmond, VA. 48 pp.

Kleopfer, J.D., J.C. Mitchell, M.J. Pinder, and S.H. Watson. 2017. **A Guide to the Snakes and Lizards of Virginia**. Virginia Department of Game and Inland Fish, Richmond, VA. 72 pp.

Kleopfer, J.D., J.C. Mitchell, P.W, Sattler, and S.H. Watson. 2020. A Guide to the Salamanders of Virginia. Virginia Department of Game and Inland Fisheries, Richmond, VA. 76 pp.

Linzey, D.W. (ed.) 1979. Endangered and Threatened Plants and Animals of Virginia. Virginia Polytechnical Institute and State University, Blacksburg, VA. 665 pp. [amphibians and reptiles on pages 375–414].

Linzey, D.W. and J. C. Michael. 1981. Snakes of Virginia. University of Virginia Press, Charlottesville, VA. xiv + 159 pp. [reprinted 1995 and 2002].

Martof, B.S., W.M. Palmer, J.R. Bailey, and J.R. Harrison. 1980. Amphibians and Reptiles of the Carolinas and Virginia. University of North Carolina Press, Chapel Hill, NC. [viii] + 264 pp. [undated softback reprint].

Mitchell, J.C. 1974. The snakes of Virginia Part I. Poisonous snakes and their look- alikes. Virginia Wildlife 35(2):16–18, 28.

Mitchell, J.C. 1974. The snakes of Virginia Part II. Harmless snakes that benefit man. Virginia Wildlife 35(4):12–15. [two versions of a combined Snakes of Virginia reprint were issued, one including 36 species and another with only 35].

Mitchell, J.C. 1975. Frogs and toads of Virginia. Virginia Wildlife 36(4):13–15, 24, 27.

Mitchell, J.C. 1976. Turtles of Virginia. Virginia Wildlife 37(6): 17–21.

Mitchell, J.C. 1977. Salamanders of Virginia. Virginia Wildlife 38(6):16–19.

Mitchell, J.C. 1977. Lizards of Virginia. Virginia Wildlife 38(8):15–16, 40.

Mitchell, J.C. 1981. Bibliography of Virginia Amphibians and Reptiles. Smithsonian Herpetological Information Service 50. 51 pp.

Mitchell, J.C. 1991. Amphibians and reptiles. Pp. 411–423 *in* K. Terwilliger (coord.). Virginia's Endangered Species. McDonald and Woodward Publishing Co., Blacksburg, VA. vii + 672 pp.

Mitchell, J.C. 1994. **The Reptiles of Virginia**. Smithsonian Institution Press, Washington, DC. xv + 352 pp.

Mitchell, J.C. 2017. Bibliography of Virginia Herpetology. Virginia Herpetological Society, Richmond, VA. 305 pp. [PDF only; https://www.virginiaherpetologicalsociety.com/bovh/biblio_va_herps_16feb17_v001.pdf].

Mitchell, J.C. and J.M. Anderson. 1994. Amphibians and Reptiles of Assateague and Chincoteague Islands. Virginia Museum of Natural History Special Publication No. 2, viii + 120 pp.

Mitchell, J.C. and C.A. Pague. 1987. A review of reptiles of special concern in Virginia. Virginia Journal of Science 38:319–328.

Mitchell, J.C. and K.K. Reay. 1999. Atlas of Amphibians and Reptiles of Virginia. Special Publication Number 1, Virginia Department of Game and Inland Fisheries, Richmond, VA. 122 pp.

Musick, J.A. 1988. The Sea Turtles of Virginia, with Notes on the Identification and Natural History. Virginia Sea Grant Program. Virginia Institute of Marine Science. Gloustcester Point, VA. 22 pp.

Pague, C.A. and J.C. Mitchell. 1987. The status of amphibians in Virginia. Virginia Journal of Science 38:304–318.

Pinder, M.J. and J.C. Mitchell. 2001. A Guide to the Snakes of Virginia. Virginia Department of Game and Inland Fish, Richmond, VA. 32 pp.

Reed, C.F. 1956. Contributions to the Herpetology of Maryland and Delmarva, 5. Bibliography to the Herpetology of Maryland, Delmarva, and the District of Columbia. Reed Herpetorium (privately published), Baltimore, MD. 9 pp.

Reed, C.F. 1956. Contributions to the Herpetology of Maryland and Delmarva, 6. An Annotated Check List of the Lizards of Maryland and Delmarva. Reed Herpetorium (privately published), Baltimore, MD. 6 pp.

Reed, C.F. 1956. Contributions to the Herpetology of Maryland and Delmarva, 7. An Annotated Check List of the Turtles of Maryland and Delmarva. Reed Herpetorium (privately published), Baltimore, MD. 11 pp.

Reed, C.F. 1956. Contributions to the Herpetology of Maryland and Delmarva, 8. An Annotated Check List of the Snakes of Maryland and Delmarva. Reed Herpetorium (privately published), Baltimore, MD. 20 pp.

Reed, C.F. 1956. Contributions to the Herpetology of Maryland and Delmarva, 9. An Annotated Check List of the Frogs and Toads of Maryland and Delmarva. Reed Herpetorium (privately published), Baltimore, MD. 19 pp.

Reed, C.F. 1956. Contributions to the Herpetology of Maryland and Delmarva, 11. An

Annotated Herpetofauna of the Del-mar-va Peninsula, including Many New and Additional Localities. Reed Herpetorium (privately published), Baltimore, MD. 11 pp. + 1 not numbered.

Reed, C.F. 1957. Contributions to the Herpetology of Maryland and Delmarva, 10. An Annotated Check List of the Salamanders of Maryland and Delmarva. Reed Herpetorium (privately published), Baltimore, MD. 6 pp.

Reed, C.F. 1957. Contributions to the herpetology of Maryland and Delmarva, 15: The herpetofauna of Somerset County, Md. Journal of the Washington Academy of Sciences 47:127–128. Reed, C.F. 1957. Contributions to the herpetofauna of Virginia, 2: The reptiles and amphibians of Northern Neck. Journal of the Washington Academy of Sciences 47:21–23.

Reed, C.F. 1957. Contributions to the herpetofauna of Virginia, 3: The herpetofauna of Accormac and Northampton Counties, Va. Journal of the Washington Academy of Sciences 47:89–91.

Richmond, N.D. 1972. Key to the lizards of Virginia. Virginia Herpetological Society Bulletin 67:2–3,13–15.

Taylor, E.A. 1958. Virginia's poisonous snakes. Virginia Wildlife 19(7):8–9.

Tobey, F.J. 1964. An aid to identification of the snakes of Virginia. Virginia Herpetological Society Bulletin 37:1–14.

Tobey, F.J. 1985. Virginia's Amphibians and Reptiles: A Distribution Survey. Virginia Herpetological Society, Richmond, WA. vi + 114 pp.

Virginia Herpetological Survey. 1968. List of Virginian amphibians and reptiles. Virginia Herpetological Society Bulletin. 56:2–6.

Virginia Herpetological Survey. 1968. Description of the turtles of Virginia. Virginia Herpetological Society Bulletin. 57:1,3–9,13–15.

Virginia Herpetological Survey. 1968. Distribution of the turtles of Virginia. Virginia Herpetological Society Bulletin. 58:1–6.

White, Jr., J.F. and A.W. White. 2002. Amphibians and Reptiles of Delmarva. Tidewater Publishers, Centreville, MD. xvi + 248 pp., 32 pp. pls.

White, Jr., J.F. and A. W. White. 2007. Amphibians and Reptiles of Delmarva. Second Edition. Delaware Nature Society, Inc., Tidewater Publishers, Centreville, Maryland. xvi + 243 pp., 32 pp. pls. [accompanied by an addendum sheet dated 2008, reprinted 2009].

Witt, W.L. 1964. Distribution of the snakes of Virginia. Virginia Herpetological Society Bulletin 38:1–6.

Wood, J.T. 1954. The distribution of poisonous snakes in Virginia. The Virginia Journal of Science 5, new series (3):152–167.

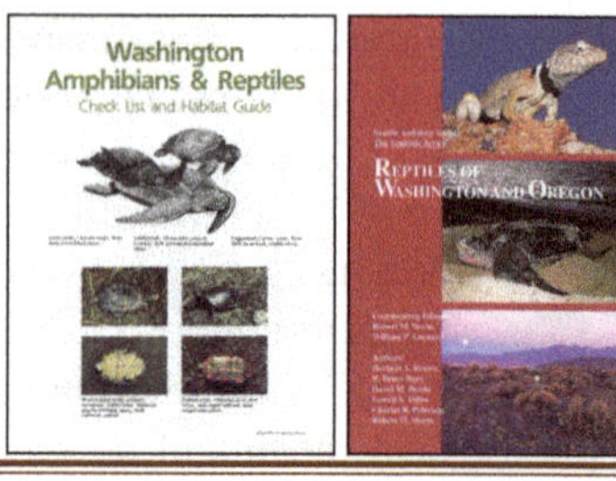

Washington

Brown, W.C. and J.R. Slater. 1939. The amphibians and reptiles of the islands of the State of Washington. Occasional Papers of Department of Biology, College of Puget Sound 4:6–31.

Carlson, D.S. 1967. A checklist of amphibians and reptiles of Washington. Bulletin of the Pacific Northwest Herpetological Society 2:7–8.

Corkran, C.C. and C. Thoms. 1996. Amphibians of Oregon, Washington, and British Columbia. Lone Pine Publishing, Edmonton, Alberta. 175 pp. [2nd edition 2006 (176 pp.), 3rd edition 2018 (176 pp.)].

Hodge, R.P. 1993. **Washington Amphibians and Reptiles: Checklist and Habitat Guide.** Washington State Game Department, Olympia, WA. 8 pp.

Johnson, M.L. 1942. A distributional checklist of the reptiles of Washington. Copeia 1942:15–18.

Johnson, M.L. 1995 (1954). Reptiles of the State of Washington. Society for Northwestern Vertebrate Biology. Northwest Fauna 3:5–80.

Leonard, W.P., H.A. Brown, L.L.C. Brown, K.R. McAllister, and R.M. Storm. 1993. Amphibians of Washington and Oregon. Seattle Audubon Society, Seattle, WA. vii + 168 pp.

McAllister, K.R. 1995. Distribution of Amphibians and Reptiles in Washington State. Northwest Fauna 3: 81–112.

Nussbaum, R.A., E.D. Brodie, Jr., and R.M. Storm. 1983. Amphibians and Reptiles of the Pacific Northwest. University Press of Idaho, Moscow, ID. 332 pp.

Owen, R.P. 1940. A list of the reptiles of Washington. Copeia 1940: 169–172.

Paulson, D. 1971. Key to Washington State reptiles and amphibians. Bulletin of the Pacific Northwest Herpetological Society 5(1):27–33.

Paulson, D. 1972. Reptiles and amphibians of Washington. Bulletin of the Pacific Northwest Herpetological Society 5(2):13–15.

Pickwell, G. 1947. Amphibians and Reptiles of the Pacific States. Stanford University Press, Stanford, CA. xiv + 236 pp. [reprinted 1949; reprinted 1972 by Dover Publications, New York, NY, xviii + 234 pp. with a new foreword and table of changes in nomenclature by G. V. Pickwell].

St. John, A. 2002. Reptiles of the Northwest. Lone Pine Press, Renton, WA. 272 pp. [revised 2021].

Slater, J.R. 1955. Distribution of Washington amphibians. Occasional Papers of Department of Biology, College of Puget Sound 16:122–154.

Slater, J.R. 1963. Key to the adult reptiles of Washington State. Occasional Papers of Department of Biology, College of Puget Sound 23:209–211.

Slater, J.R. 1963. Distribution of Washington reptiles. Occasional Papers of Department of Biology, College of Puget Sound 24:212–232.

Slater, J.R. 1964. Key to the adult amphibians of Washington State. Occasional Papers of

Department of Biology, College of Puget Sound 25:235–236.

Slater, J.R. and W.C. Brown. 1941. Island records of the amphibians and reptiles for Washington. Occasional Papers of Department of Biology, College of Puget Sound 13:74–77.

Slevin, J.A. 1934. A Handbook of Reptiles and Amphibians of the Pacific States. California Academy of Sciences, San Francisco, CA. 73 pp.

Storm, R.M. and W.P. Leonard (eds). 1995. **Reptiles of Washington and Oregon**. Seattle Audubon Society, Seattle, WA. vii + 176 pp.

Wagner, E. 1976. The turtles of Washington State. Pacific Search 10(6):28–29.

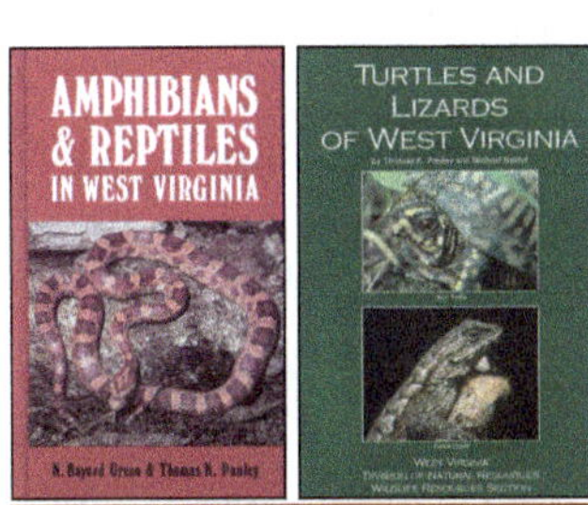

West Virginia

Czarnowsky, J. R. 1980. A Guide to the Amphibians and Reptiles of the Cooper's Cove Area (Hampshire County, W. Va.). Privately published. [vi] + 110 pp. (+ 1 loose p. correction to maps).

Green, N. B. 1937. The amphibians and reptiles of Randolph County, West Virginia. Herpetologica 1:113–116.

Green, N. B. 1943. The snakes of West Virginia. West Virginia Conservation 6(March):7, 16–18.

Green, N. B. 1943. The salamanders of West Virginia. West Virginia Conservation 6(May):12–13, 18–19.

Green, N. B. 1948. Toads, frogs and turtles of West Virginia. West Virginia Conservation. 11(October):11–12.

Green, N. B. 1948. The herpetological collections of the West Virginia Biological Survey. Proceedings of West Virginia Academy of Sciences 20:57–64.

Green, N. B. 1948. An Annotated List of Amphibians and Reptiles Known to Occur in West Virginia. West Virginia Conservation Commission, Charleston, WV. 12 pp.

Green, N. B. 1951. Key to the Amphibians and Reptiles of West Virginia. Marshall College, Huntington, WV. 26 pp., 4 pls.

Green, N. B. 1969. The occurrence and distribution of turtles in West Virginia. Proceedings of West Virginia Academy of Sciences 41:1–14.

Green, N. B. 1954. Amphibians and Reptiles in West Virginia: Their Identification and Distribution. Marshall University, Huntington, WV. 38 pp., 6 pls. 47 pp. [revised 1963 (38 pp., 6 pls.), 1964 ([ii] + 47 pp.), 1969 ([ii] + 47 pp.), and possibly others].

Green, N. B. 1976. The Amphibians of West Virginia. West Virginia Encyclopedia 1:73–85.

Green, N. B. 1976. The Reptiles of West Virginia. West Virginia Encyclopedia 18:4015–4029.

Green, N. B, and B. Dowler. 1967. Amphibians and reptiles of the Little Kanawha River Basin. Proceedings of West Virginia Academy of Sciences 38:50–57.

Green, N. B., F. Jernejcic, T.K. Pauley, and D. Pursley. 1995. Snakes of West Virginia. West Virginia Nongame Wildlife and Natural Heritage Program, Elkins, WV. 14 pp. [revised 2002 (19 pp.), reprinted 2006].

Green, N. B. and D. Pursley. 1975. West Virginia snakes. Wonderful West Virginia 39:2–3,6–7,26.

Green, N. B., and T.K. Pauley. 1987. **Amphibians and Reptiles in West Virginia**. University of Pittsburgh Press, Pittsburgh, PA. xi + 241 pp.

Netting, M. G. 1933. The amphibians of West Virginia. Part I. Salamanders. West Virginia Wildlife 11(3–4):5–6,15.

Netting, M. G. 1933. The amphibians of West Virginia. Part II. Toads and frogs. West Virginia Wildlife 11(5–6):4–5.

Pauley, T.K. 1996. Toads and Frogs of West Virginia. West Virginia Nongame Wildlife and Natural Heritage Program, Elkins, WV. 12 pp.

Pauley, T.K. 2001. Toads and Frogs of West Virginia. West Virginia Natural Resources, Elkins, WV. 12 pp.

Pauley, T.K. 2004. Salamanders of West Virginia. West Virginia Natural Resources, Elkins, WV. 20 pp.

Pauley, T.K. 2011. Toads and Frogs of West Virginia, Second Edition. West Virginia Natural Resources, Elkins, WV. 13 pp.

Pauley, T. K. 2012. Snakes. West Virginia Division of Natural Resources, Wildlife Resources Section, Elkins, West Virginia. 20 + [1] pp.

Pauley, T.K. 2013. Salamanders of West Virginia. West Virginia Natural Resources, Elkins, WV. 20 pp.

Pauley, T.K. and Michael Seidel. 1996. Turtles and Lizards of West Virginia. West Virginia Nongame Wildlife and Natural Heritage Program, Elkins, WV. 12 pp.

Pauley, T.K. and Michael Seidel. 2002. **Turtles and Lizards of West Virginia**. West Virginia Natural Resources, Elkins, WV. 15 pp.

Pauley, T.K. and Michael Seidel. 2014. Turtles and Lizards of West Virginia. West Virginia Natural Resources, Elkins, WV. 17 pp.

Richmond, N.D. and G. Boggess. 1941. A key to the reptiles and amphibians of West Virginia. West Virginia University Bulletin 42:1–9.

Venable, N.J. 2001. Scales, Skinks, Scutes & Newts. An Introduction to West Virginia's Amphibians and Reptiles. West Virginia University Extension Service, Morgantown, WV. 59 pp

Wisconsin

Carr, C.F. 1890. Salamanders. Wisconsin Naturalist 1:54–56.

Casper, G.S. 1985. Wisconsin's harmless snakes. Lore 35(2):10–19.

Casper, G.S. 1996. Geographic Distributions of the Amphibians and Reptiles of Wisconsin. Milwaukee Public Museum, Milwaukee, WI. 87 pp.

Casper, G.S. 1998. Review of the Status of Wisconsin Amphibians pp. 199–205 *in* Lannoo, Michael J. (ed.) Status and Conservation of Midwestern Amphibians. University of Iowa Press, Iowa City, IA. xviii + 507 pp.

Christoffel, R., R. Hay, and M. Monroe. 2002. Turtles and Lizards of Wisconsin. Wisconsin Department of Natural Resources, Madison, WI. 48 pp.

Christoffel, R., R. Hay, and L. Ramirez. 2000. Snakes of Wisconsin. Wisconsin Department of Natural Resources, Madison, WI. 32 pp.

Christoffel, R., R. Hay, R. Paloski and L. Ramirez. 2008. Snakes of Wisconsin, Second Edition. Wisconsin Department of Natural Resources, Madison, WI. 32 pp.

Christoffel, R., R. Hay, and M. Wolfgram. 2001. Amphibians of Wisconsin. Wisconsin Department of Natural Resources, Madison, WI. 44 pp.

Christoffel, R., R. Hay, R. Paloski, and M. Wolfgram. 2009. Amphibians of Wisconsin, Second Edition. Wisconsin Department of Natural Resources, Madison, WI. 44 pp.

Craven, S.C. and G.J. Knutson. 1982. Snakes of Wisconsin. University of Wisconsin – Extension, Madison, WI. 8 pp.

Dickinson, W.E. 1949. Field Guide to the Lizards and Snakes of Wisconsin. Milwaukee Public Museum. Popular Science Handbook Series No. 2. 70 pp.

Dickinson, W.E. 1965. **Handbook of Amphibians and Turtles of Wisconsin**. Milwaukee Public Museum. Popular Science Series No. 10. 45 pp.

Dickinson, W.E. 1972. The amphibians and reptiles of Forest, Florence, and Marinette counties with special reference to the Pine, Popple, and Pike Watersheds. Transactions of the Wisconsin Academy of Science, Arts and Letters 60:303–308.

Dlutkowski, L.A., P.A. Cochran, and M.J. Mossman. 1987. Bibliography of Wisconsin Herpetology. Department of Natural Resources, Wisconsin Endangered Resources Report 28. 28 pp.

Higley, W.K. 1889. Reptiles and batrachians of Wisconsin. Transactions of the Wisconsin Academy of Science, Arts and Letters 7:155–176.

Hoy, P.R. 1883. Catalogue of the cold–blooded vertebrates of Wisconsin. Geological Survey of Wisconsin 1:422–426.

Kapfer, J.M. and D.J. Brown (eds). 2022. **Amphibians and Reptiles of Wisconsin**. University of Wisconsin Press, Madison, WI. xxi +1175 pp.

Korb, R.M. 2001. Wisconsin Frogs. Places to Hear Frogs and Toads Near Our Urban Areas. Northeastern Wisconsin Audubon Inc., Green Bay, WI. 80 pp. + CD.

Parmelee, J.R., M.G. Knutson, and J.E. Lyon. 2002. A Field Guide to Amphibian Larvae and Eggs of Minnesota, Wisconsin, and Iowa. U.S. Geological Survey Information and Technology Report 2002–0004. 38 pp.

Pope, T.E.B. and W.E. Dickinson. 1928. The Amphibians and Reptiles of Wisconsin. Bulletin of the Public Museum of the City of Milwaukee 8:1–138.

Sheldon, A.B. 1981. The snakes of Wisconsin. Wisconsin Sportsman 10(4):56–60.

Sheldon, A.B. 1991. The turtles of Wisconsin. Wisconsin Outdoor Journal 5(2):53–56.

Sheldon, A.B. 2006. Amphibians and Reptiles of the North Woods. Kollath and Stensaas, Duluth, MN. 148 pp.

Sheldon, A.B. 2021. Amphibians and Reptiles of Minnesota, Wisconsin, and Michigan. Kollath and Stensaas, Duluth, MN. 201 pp.

Sheldon, A.B. and D. Nedrelo. 1986. The lizards of Wisconsin. Wisconsin Sportsman

15(5):56–57.

Tekiela, S. 2004. Reptiles and Amphibians of Wisconsin Field Guide. Adventure Publications, Cambridge, MN. xxiv + 147 pp. + CD.

Tekiela, S. 2014. Reptiles and Amphibians of Minnesota, Wisconsin and Michigan. Adventure Publications, Cambridge, MN. 216 pp.

Vogt, R.C. 1981. Natural History of Amphibians and Reptiles of Wisconsin. Milwaukee Public Museum, Milwaukee, WI. 205 pp.

Watermolen, D. J. 1992. Wisconsin Herpetology: A Bibliographic Update with Taxonomic and Geographic Indices. Wisconsin Endangered Resources Report No. 87. 13 pp.

Wyoming

Baxter, G.T. 1947. **The amphibians and reptiles of Wyoming**. Wyoming Wildlife 11:30–34.

Baxter, G.T, and M.D. Stone. 1980. Amphibians and Reptiles of Wyoming. Wyoming Game and Fish Department, Cheyenne, WY. vi + 137 pp. [2nd edition1985].

Carpenter, C.C. 1953. An ecological survey of the herpetofauna of the Grand Teton – Jackson Hole area of Wyoming. Copeia 1953:170–174.

Corn, P.S., R. Bruce Bury, and H.H. Welsh. 1984. Selected Bibliography of Wyoming Amphibians and Reptiles. Smithsonian Herpetological Information Service 59. 34 pp.

Koch, E.D. and C.R. Peterson. 1995. Amphibians and Reptiles of Yellowstone and Grand Teton National Parks. University of Utah Press, Salt Lake City, UT. xviii + 188 pp.

Lewis, D. 2011. **A Field Guide to the Amphibians and Reptiles of Wyoming**. Wyoming Naturalist, Douglas, WY. vii +183 pp. [also published in same year as the Armchair Guide to the Amphibians and Reptiles of Wyoming with bibliographic details and layout identical, but larger page size].

Livo, L.J. 1995. Identification Guide to Montane Amphibians of Southern Rocky Mountains. Colorado Division of Wildlife, Denver, CO. 25 pp.

Parker, J. and S. Anderson. 2001. Identification Guide to the Herptiles of Wyoming. Wyoming Game and Fish, Cheyenne, WY. 40 pp.

Turner, F.B. 1955, Reptiles and Amphibians of Yellowstone Park. Yellowstone National Park, Yellowstone Interpretive Series No. 5. vi + 40 pp.

Alberta

Anonymous. 2010. **Reptiles of Alberta**. Alberta Conservation Association, Sherwood Park, AB. 11 pp.

Harper, F. 1931. Amphibians and reptiles of the Athabasca and Great Slave Lakes region. Canadian Field-Naturalist 45:68–70.

Hodge, R.P. 1972. Amphibians of Alberta. Alberta Conservationist. Summer 1972:11–14.

Lewin, V. 1963. The herpetofauna of Southeastern Alberta. Canadian Field-Naturalist 77:203–214.

Preble, E.A. 1908. Reptiles and amphibians of the Athabaska-MacKenzie Region. North American Fauna 27:500–502.

Russell, A.P. and A.M. Bauer. 1991. The amphibians and reptiles of the Calgary area. Pica 11:3–15. [also published in the same year in Alberta Naturalist 21:123–128 and Alberta Reptile and Amphibian Society 26:12–23 without authorization]

Russell, A.P. and A.M. Bauer. 1993. The Amphibians and Reptiles of Alberta. University of Calgary Press, Calgary, AB. x + 264 pp.

Russell, A.P. and A.M. Bauer. 2000. **The Amphibians and Reptiles of Alberta – Revised Edition**. University of Calgary Press, Calgary, AB. xii + 279 pp.

British Columbia

Carl, G. C. 1943. **The Amphibians of British Columbia**. British Columbia Provincial Museum Handbook No. 2. 62 pp. [2nd edition 1951(62 pp.), 3rd edition 1959 (62 pp.), 4th edition 1966 (63 pp.), 5th edition 1973 (63 pp.)].

Carl, G. C. 1944. The Reptiles of British Columbia. British Columbia Provincial Museum Handbook No. 3. 60 pp. [2nd edition 1951 (65 pp), 3rd edition1960 (65 pp.), reprinted 1968].

Corkran, C.C. and C. Thoms. 1996. Amphibians of Oregon, Washington, and British Columbia. Lone Pine Publishing, Edmonton, Alberta. 175 pp. [2nd edition 2006 (176 pp.), 3rd edition 2018 (176 pp.)].

Cowan, I. McT. 1937. A review of the reptiles and amphibians of British Columbia. Report of British Columbia Provincial Museum of Natural History 1936:16–25.

Green, D.M. 1999. The Amphibians of British Columbia: A Taxonomic Catalogue. Wildlife Branch and Resources Inventory Branch, British Columbia Ministry of Environment, Lands and Parks, Wildlife Bulletin No. B–87. viii + 22 pp.

Green, D.M., and R.W. Campbell. 1984. The Amphibians of British Columbia. British Columbia Provincial Museum Handbook 45. viii + 101 pp.

Gregory, P.T. and R.W. Campbell. 1984. The Reptiles of British Columbia. British Columbia Provincial Museum Handbook 44. viii + 103 pp.

Gregory, L.A. and Gregory, P.T. 1999. The Reptiles of British Columbia: A Taxonomic Catalogue. Wildlife Branch and Resources Inventory Branch, British Columbia Ministry of Environment, Lands and Parks, Wildlife Bulletin No. B–88. viii + 27 pp.

Hardy, G.A. 1926. Amphibians of British Columbia. Report of Provincial Museum of Natural History 1925:21–24.

Logier, E.B.S. 1932. Some account of the amphibians and reptiles of British Columbia. Transactions of the Royal Canadian Institute 18:311–336.

Matsuda, B.M., D.M. Green, and P.T. Gregory. 2006. **Amphibians and Reptiles of British Columbia**. Royal BC Museum, Victoria, BC. 266 pp.

Orrico, L. 1957. British Columbia turtles. Victoria Naturalist 14:11–12.

Ovaska, K., and P. Govindarajulu. 2010. A Guide to Amphibians of British Columbia North of 50°. British Columbia Ministry of Environment, Vancouver, BC. 14 pp.

St. John, A. 2002. Reptiles of the Northwest. Lone Pine Press, Renton, WA. 272 pp. [revised 2021].

Taylor, J.M. 1979. Amphibians, Reptiles, and Mammals of British Columbia. Department of Zoology, University of British Columbia, Vancouver, BC. 144 pp.

Manitoba

Harper, J. 1963. Amphibians and reptiles of Keewatin and north Manitoba. Proceedings of the Biological Society of Washington 76:159–168.

Jackson, V.W. 1934. **A Manual of Vertebrates of Manitoba**. University of Manitoba, Winnipeg, MB. 41 pp. [amphibians and reptiles on pp. 9–10].

Preble, E.A. 1902. A biological investigation of the Hudson Bay region. North American Fauna No. 22. 140 pp., 14 pls., 1 folding map. [amphibians and reptiles on pp. 133– 134] .

Preston, W.B. 1974. Amphibians and Reptiles. Pp. 73–96 *in* Wrigley, Robert E. (ed.). Animals of Manitoba. Manitoba Museum of Man and Nature, Winnipeg, MB. 158 pp.

Preston, W.B. 1982 **The Amphibians and Reptiles of Manitoba**. Manitoba Museum of Man and Nature, Winnipeg, MB. 128 pp.

Seton, E.T. 1918. A list of the turtles, snakes, and batrachians of Manitoba. Ottawa Naturalist 32:79–83.

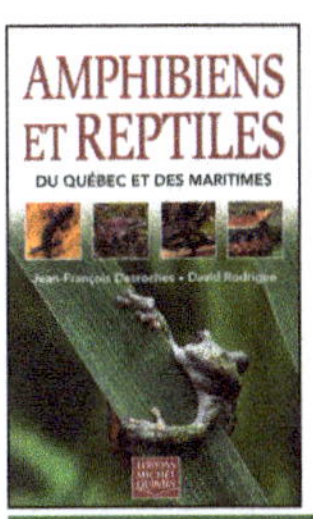

Maritimes

Cox, P. 1899. Freshwater fishes and Batrachia of the Peninsula of Gaspé. P.Q. and their distribution in the Maritime Provinces of Canada. Transactions of the Royal Society of Canada 5(4):141–154.

Cox, P. 1899. A preliminary list of the Batrachia of the Peninsula of Gaspé. P.Q. and the Maritime Provinces of Canada. Ottawa Naturalist 13:194–195.

Cox, P. 1903. The snakes of the Maritime Provinces of Canada. Proceedings of Miramichi Natural History Association 3:11–20.

Desroches, J.F. and D. Rodrique. 2004. **Amphibians et Reptiles du Québec et des Maritimes**. Editions Michel Quintin, Waterloo, QC. 288 pp.

Rodrique, D. and J.F. Desroches. 2018. Amphibiens et Reptiles du Québec et Des Maritimes. Editions Michel Quintin, Waterloo, QC. 376 pp.

New Brunswick

Cox, P. 1898. Batrachia of New Brunswick. Bulletin of Natural History Society of New Brunswick 4:64–66.

Cox, P. 1899. Anoura of New Brunswick. Proceedings of Miramichi Natural History Association 1899(1):9–19.

Gorham, S.W. 1970. **The Amphibians and Reptiles of New Brunswick**. New Brunswick Museum Monographic Series No. 6, Saint John, NB. ix + 30 pp.

Newfoundland

Maret, E. 1867. Frogs of Newfoundland. Proceedings of the Nova Scotia Institute of Science 1(3):6.

Maunder, J.E. 1983. **Amphibians of the Province of Newfoundland**. Canadian Field-Naturalist 97:33–46.

Northwest Territories

Environment and Natural Resources [Carrière, S., P. Lacroix, J. Stewart, and J. Wilson]. 2021. Field Guide to Amphibians and Reptiles of the Northwest Territories. Environment and Natural Resources, Government of the Northwest Territories, Yellowknife, NT. 58 pp.

Harper, F. 1931. Amphibians and reptiles of the Athabasca and Great Slave Lakes region. Canadian Field-Naturalist 45:68–70.

Hodge, R.P. 1971. Herpetology north of 60°. The Beaver, Summer 1971:36–38.

Hodge, R.P. 1976. **Amphibians and Reptiles in Alaska, The Yukon and Northwest Territories**. Alaska Northwest Publishing, Anchorage, AK. Xi + 89 pp. [reptinted 1977].

Larsen, K.W. and Gregory, P.T. 1988. **Amphibians and reptiles in the Northwest Territories**. Occasional Papers of the Prince of Wales Northern Heritage Centre 3:31–51.

Nova Scotia

Bleakney, J.S. 1952. The amphibians and reptiles of Nova Scotia. Canadian Field-Naturalist 66: 125–129.

Bleakney, J.S. 1963. Notes on the distribution and life histories of turtles in Nova Scotia. Canadian Field-Naturalist 77:67–76.

Gilhen, J. 1966. **Tips on Turtles**. Nova Scotia Museum, Halifax. 6 pp.

Gilhen, J. 1984. **Amphibians and Reptiles of Nova Scotia**, Nova Scotia Museum, Halifax, NS. 162 pp.

Gilhen, J. and F. Scott. 1981. Distribution, Habitats, and Vulnerability of Amphibians, Reptiles and Small Mammals in Nova Scotia. Nova Scotia Museum Curatorial Report No. 45. 21 pp.

Gilpin, J.B. 1875. On the serpents of Nova Scotia. Transactions of Nova Scotia Institute of Science 4:80–88.

MacKay, A.H. 1896. Batrachia and Reptilia of Nova Scotia. Proceedings and Transactions of Nova Scotia Institute of Science 9(2): xliii.

Martin, J.L. 1969. The Amphibians and Reptiles of Nova Scotia. Nova Scotia Museum, Halifax, NS. 17 pp.

HERPETOLOGY NORTH OF 60

Nunavut

Harper, J. 1963. **Amphibians and reptiles of Keewatin and north Manitoba**. Proceedings of the Biological Society of Washington 76:159–168.

Hodge, R.P. 1971. **Herpetology north of 60°**. The Beaver, Summer 1971:36–38.

Hodge, R.P. 1976. Amphibians and Reptiles in Alaska, The Yukon and Northwest Territories. Alaska Northwest Publishing, Anchorage, AK. Xi + 89 pp. [reptinted 1977].

Preble, E.A. 1902. A biological investigation of the Hudson Bay region. North American Fauna No. 22. 140 pp., 14 pls., 1 folding map. [amphibians and reptiles on pp. 133–134].

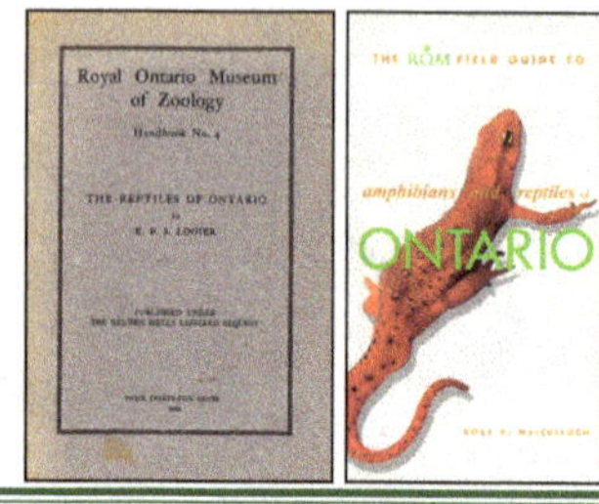

Ontario

Anonymous [R.C. McCabe]. 1958. The Reptiles of Algonquin Park. Ontario Department of Lands and Forests, [Algonquin Park] Museum Bulletin 1. 22 pp. [revised edition 1958 ([ii] + 19 pp.)].

Bosco, T., Johnson, B., Karch, M., Lathrop, A., Mason, T., McDonald, K., Phillips, J., Prior, P., Richards, N., Salvatori, N., Snow, K., Urquhart, J, and Viggiani, P. 2012. Reptiles and Amphibians of Toronto. City of Toronto, Toronto, ON. 72 pp.

Brooks, R.J., D. Strickland, and R.J. Rutter, 2003. Reptiles and Amphibians of Algonquin Provincial Park. Whitney, ON, The Friends of Algonquin Park in cooperation with Ontario Ministry of Natural Resources. 48 pp.

Christie, P. 1997. Reptiles and Amphibians of Prince Edward County, Ontario. Natural Heritage Books, Toronto. 143 pp.

Desroches, J.-F.., F. Schuler, I. Picard, and L.-P. Gagnon. 2010. A Herpetological survey of the James Bay Area of Québec and Ontario. Canadian Field-Naturalist 124:299–315.

Froom, B. 1967. Ontario Snakes. Department of Lands and Forests, Ontario. Toronto, ON. 36 pp. [reprinted 1971].

Froom, B. 1971. Ontario Turtles. Department of Lands and Forests, Ontario. Toronto, ON. 25 pp. [revised 1975].

Garnier, J.H. 1881. List of Reptilia of Ontario. The Canadian Sportsman and Naturalist 1(5):37–39.

Gillingwater, S. D. and A.S. MacKenzie. 2015. Photo Field Guide to the Reptiles and Amphibians of Ontario. St. Thomas Field Naturalist Club, St. Thomas, ON. 144 pp.

Johnson, B. 1989. Familiar Amphibians and Reptiles of Ontario. Natural Heritage/ Natural History Inc., Toronto, ON. 168 pp.

Logier, E.B.S. 1937. The Amphibians of Ontario. Royal Ontario Museum of Zoology, Handbook No. 3. Toronto, ON. 16 pp.

Logier, E.B.S. 1939. **The Reptiles of Ontario**. Royal Ontario Museum of Zoology, Handbook No. 4. Toronto, ON. ii + 63 pp. + 7 pls.

Logier, E.B.S. 1958. The Snakes of Ontario. University of Toronto Press, Toronto, ON. x + 94 pp.

MacCulloch, R.D. 2002. **The ROM Field Guide to Amphibians and Reptiles of Ontario** Royal Ontario Museum. Toronto, ON. 168 pp.

McBride, B. 1969. The Turtles of Ontario. Royal Ontario Museum Information Leaflet, Toronto, ON.

Mills, P.S. 2016. Metamorphis: Ontario's Amphibians at All Stages of Development. Bog Hunter, Brampton, ON. 104 pp.

Milnes, H. 1946. Amphibians and reptiles of Oxford County, Ontario. The Canadian

Field- Naturalist 60:1–4.

Nash, C.W. 1905. Batrachians and reptiles of Ontario. pp. 5–18 *in* Check List of the Vertebrates of Ontario and Catalogue of Specimens in the Biological Section of the Provincial Museum. Batrachians, Reptiles, Mammals. Ontario Department of Education, Toronto, ON. [reprinted with original pagination in Nash, C.W. 1908. Manual of Vertebrates of Ontario. Ontario Department of Education, Toronto, ON. 235 pp.].

Parsons, H. 1976. Foul and Loathsome Creatures. Minister of Supply and Services Canada, Ottawa, ON. [viii] + 58 pp., 8 pls. [French translation published as Ces Horribles et Dégoutantes Créatures. [viii] + 63, 9 pls.].

Perkins Bull, W. 1938. From Amphibians to Reptiles, Shy Swamp-Dwellers in Study, Picture and Legend. The Perkins Bull Foundation/George J. McLeod Ltd., Toronto, ON. 89 + [vi], 9 pls., 1 folding map, 1 folding chart.

Peters, M. G. 2004. Life-Size Amphibians and Reptiles of Southern Ontario. Orders: Anura, Caudata, Squamata and Testudines. Privately Published. Guelph, Ontario. v + 53 pp.

Piersol, W.H. 1913. Amphibia. Pp. 242–248 in Faull, J.H. (ed.), The Natural History of the Toronto Region, Ontario, Canada. The Canadian Institute, Toronto, ON.

Rowell, J. 2012. The Snakes of Ontario. Art BookBindery, Winnipeg, MB. vi + 411 pp.

Strickland, D. and Rutter, R.J. 1976. Reptiles and Amphibians of Algonquin Provincial Park. Whitney, ON, The Friends of Algonquin Park in cooperation with Ontario Ministry of Natural Resources. 32 pp. [unnumbered], [revised in 1978, reprinted in 1986, 1992, 2000].

Toner, G.C. and N. de St. Remy. 1941. Amphibians of eastern Ontario. Copeia 1941:10–13.

Williams, J.B. 1913. Reptiles. Pp. 238–241 in Faull, J.H. (ed.), The Natural History of the Toronto Region, Ontario, Canada. The Canadian Institute, Toronto.

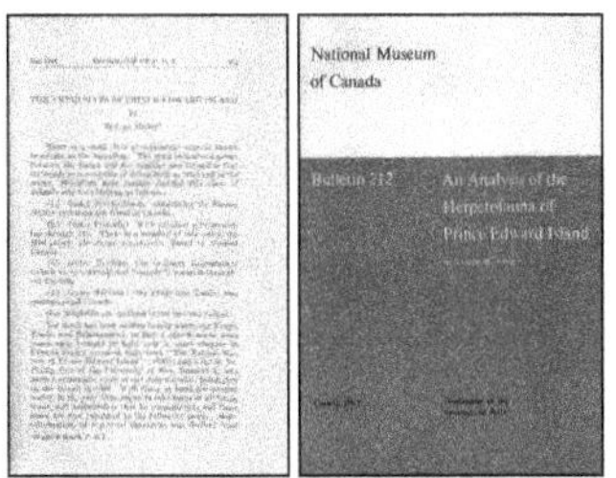

Prince Edward Island

Cook, Francis R. 1967. **An Analysis of the Herpetofauna of Prince Edward Island.** Natural Museum Canada Bulletin 212. 60 pp.

Hurst, B. 1944. **The amphibians of Prince Edward Island**. Acadian Naturalist 1(3):111–116.

Mellish, J.T. 1877. Notes on the serpents of Prince Edward Island. Transactions of the Nova Scotia Institute of Science 4(2):163–167.

Stewart, J. 1806. An Account of Prince Edward Island in the Gulph of St. Lawrence, North America. Winchester and Son, London. xiii + [2] + 304 pp.,1 folding map. [amphibians and reptiles on pp. 78–79].

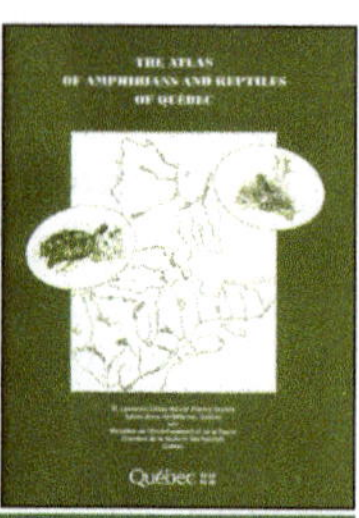

Québec

Alexandre, [Frère]. 1937. Les couleuvres du Québec. Société Canadienne d'Histoire Naturelle, Bibliothèque des Jeunes Naturalistes, Tract 26. 4 pp. [reprinted 1938 in Bibliothèque des Jeunes Naturalistes, Tracts Nos. 1–50. Société Canadienne d'Histoire Naturelle. Montréal, QC].

Alexandre, [Frère]. 1937. Les tortues du Québec. Société Canadienne d'Histoire Naturelle, Bibliothèque des Jeunes Naturalistes, Tract 39. 4 pp. [reprinted 1938 in Bibliothèque des Jeunes Naturalistes, Tracts Nos. 1–50. Société Canadienne d'Histoire Naturelle. Montréal, QC].

Alexandre [Frère]. 1945. Nos grenouilles et nos crapauds. Société Canadienne d'Histoire Naturelle, Montréal, QC. 4 pp. Bibliothèque des Jeunes Naturalistes,, Tract 82.

Bider, J. R., and S. Matte. 1996. **The Atlas of Amphibians and Reptiles of Quebéc**. St. Lawrence Valley Natural History Society and Ministére de l'Environment et de la Faune du Quebéc, Direction de la Faune et des Habitats. Quebéc, QC. v + 106 pp.

Cox, P. 1899. Freshwater fishes and Batrachia of the Peninsula of Gaspé. P.Q. and their distribution in the Maritime Provinces of Canada. Transactions of the Royal Society of Canada 5(4):141–154.

Cox, P. 1899. A preliminary list of the Batrachia of the Peninsula of Gaspé. P.Q. and the Maritime Provinces of Canada. Ottawa Naturalist 13:194–195.

Desroches, J.–F. and D. Rodrique. 2004. Amphibians et Reptiles du Québec et des Maritimes. Editions Michel Quintin, Waterloo, QC. 288 pp.

Desroches, J.–F., F. Schuler, I. Picard, and L.–P. Gagnon. 2010. A Herpetological survey of the James Bay area of Québec and Ontario. Canadian Field- Naturalist 124:299–315.

Harper, F. 1956. Amphibians and reptiles of the Ungava Peninsula. Proceedings of the Biological Society of Washington 69:93–104.

Mélançon, C. 1950. **Inconnus et Méconnus (Amphibiens et Reptiles de la Province de Québec)**. La Société zoologique de Québec, Inc., Québec, QC. 148 pp. + 2 pp. Index. [2nd edition 1961].

Rodrique, D. and J.–F. Desroches. 2018. Amphibiens et Reptiles du Québec et Des Maritimes. Editions Michel Quintin, Waterloo, QC. 376 pp.

Trapido, H. and Clausen, R.T. 1938. Amphibians and reptiles of eastern Québec. Copeia 1938:117–125.

Vladykov, V.D. 1941. Preliminary list of Amphibia from the Laurentides Park in the Province of Québec. Canadian Field-Naturalist 55:83–84.

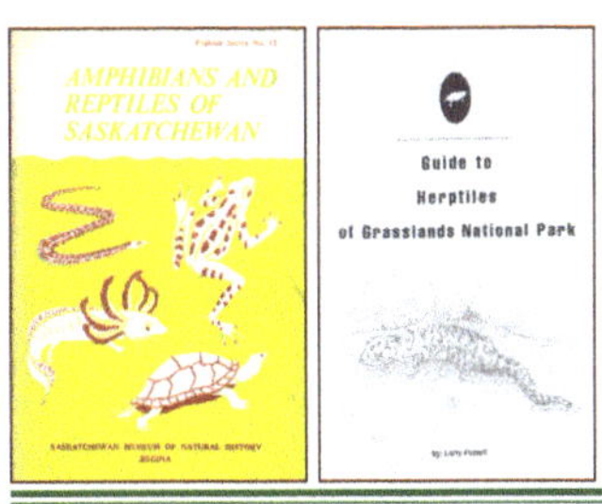

Saskatchewan

Cook, F.R. 1965. Additions to the known range of some Amphibians and reptiles in Saskatchewan. Canadian Field-Naturalist. 79 (2):112–120

Cook, F.R. 1966. **A Guide to the Amphibians and Reptiles of Saskatchewan**. Saskatchewan Museum of Natural History, Popular Series 13, 40 pp. [reprinted 1970, 1977, revised 1978 (28 pp.)].

Hooper, D.F. 1972. Turtles, snakes and salamanders of east-central Saskatchewan. Blue Jay 50:72–75.

Powell, L. n.d. [2005]. **Guide to the Herptiles of Grasslands National Park**. Prairie Wind and Silver Sage – Friends of Grasslands Inc. [iv] + 43 pp.

Secoy, D.M. 1987. Status report on the reptiles and amphibians of Saskatchewan. Pp. 139–141 *in* G. Holroyd (ed.) Endangered species in the Prairie Provinces. Provincial Museum of Alberta, Edmonton, AB

Yukon

Anonymous. 2005. **Yukon Amphibians**. Government of Yukon, Whitehorse, YT. 22 pp.

Hodge, R.P. 1971. Herpetology north of 60°. The Beaver, Summer 1971:36–38.

Hodge, R.P. 1976. Amphibians and Reptiles in Alaska, The Yukon and Northwest Territories. Alaska Northwest Publishing, Anchorage, AK. xi + 89 pp. [reprinted 1977].

International Society for the History and Bibliography of Herpetology

The ISHBH publishes the journal *Bibliotheca Herpetologica*, a peer-review journal containing articles, essays, bibliographies, and biographies. All issues of *Bibliotheca Herpetologica* are available Open Access on the society website, www.ISHBH.com. A print copy of each volume of *Bibliotheca Herpetologica*, containing all articles published that year, is distributed to current members.

The name of the journal, up to volume 5(1), was: *International Society for the History and Bibliography of Herpetology Newsletter and Bulletin*. All of these issues are also available, Open Access on the society's website.

Wahlgreniana. A series of book-length works complementing the ISHBH journal, *Bibliotheca Herpetologica*. Books in this series are published on an irregular basis and are sold separately from ISHBH subscriptions, discounted for society members. *Wahlgreniana* is named in honor of Richard Wahlgren (1946–2019) A founding member and first Chairman of the International Society for the History and Bibliography of Herpetology. Without Richard's tireless dedication to ISHBH, the society could not have made it through its early years.

Volume 1—May 2022

Bour, Roger and Josef F. Schmidtler. 2022. *Nikolaus Michael Oppel's Drawings, Watercolors, and Engravings 3. Crocodiles (1807–1817): A comparative study of some historical and recent crocodile illustrations*. ISHBH, Salt Lake City, x, 184 p. Hardcover, ISBN 978-0-578-29399-8. Includes a complete facsimile of Tiedemann, F., M. Oppel & J. Liboschitz 1817. *Naturgeschichte der Amphibien. Erstes Heft. Gattung Krokodil.* Joseph Engelmann, Heidelberg. v–vi, 1–88, vii–viii. Auf Kosten der Verfasser, München, 15 pls. Retail: $65.00; ISHBH members $39.00. Postage is additional.

Volume 2—October 2022

Dodd, C. Kenneth, Jr. 2022. *Bibliography of the Anurans of the United States and Canada. Version 3. Part 1: 1698–2012. Part 2: 2013-2021*. ISHBH, Salt Lake City, x, 282 p. Hardcover, ISBN: 979-8-218-06245-3; eBook ISBH: 979-8-218-06246-0. Retail: $35.00; ISHBH members $21.00. Postage is additional.

Volume 3—June 2024

Moriarty, John J. and Aaron M. Bauer. 2024. *State and Provincial Amphibian and Reptile Publications For the United States and Canada, Second Edition*. ISHBH, Salt Lake City, vi., 85 pp. Paperback, ISBN: 979-8-218-44771-7. Retail: $20.00; ISHBH members $12.00. Postage is additional.

Join the society & order publications at: www.ISHBH.com

www.ingramcontent.com/pod-product-compliance
Lightning Source LLC
Chambersburg PA
CBHW060254030426
42335CB00014B/1694
9798218447717